HEALING FROM NARCISSISTIC MOTHERS:

THE DEFINITIVE "2 IN 1" GUIDE FOR DAUGHTERS

Two Manuscripts:
NARCISSISTIC MOTHERS
SELF-LOVE WORKBOOK FOR WOMEN

DESIRÉE SHANNON

Copyright © 2019 Desirée Shannon

All rights reserved

Table of Content

BOOK 1: NARCISSISTIC MOTHERS

Introduction 1

Chapter 1

What Does Narcissistic Personality Disorder Look Like? 7

What Causes Narcissistic Personality Disorder? 10

Are There Any Cures for Narcissistic Personality Disorder? 11

Chapter 2

A Mother That is Threatened by Her Child 13

An Effort at Self-Fulfillment Through You 14

The Development of a Superficial Image 16

Chapter 3

You May Be Chronically Ashamed of Yourself 17

Childhood Abuse May Lead to Adulthood Abuse or Toxic Relationship Patterns 19

You May Reflect Some of Your Mother's Symptoms 20

There May Be the Feeling of a Deep Void in Your Life 21

Chapter 4

What to Do If Your Relationship Must End Completely 24

What to Do If You Need to Minimize Your Relationship 25

What to Do If You Need to Stay Consistent in Your Relationship 26

Chapter 5
- You Live Only to Serve Others 30
- You Cannot Do Anything Right 30
- Your Perception of Reality is Not True 31

Chapter 6
- Understand the Cycle of Narcissism 36
- Surround Yourself with Supportive Outsiders 37
- Prepare Yourself For The Emotional Experience 39

Chapter 7
- Start Viewing Your Situation for What it Truly Is 42
- Explain the Story from An Outsider's Perspective 43
- Keep the Reality in Mind When Making Your Exit 44

Chapter 8
- Start to Get to Know Yourself in A Deeper Way 48
- Keep Your Promises to Yourself, Always 49
- Learn to Tell Yourself the Things You Need to Hear 50
- Stop Judging Yourself and Your Feelings 50
- Practice Parenting Your Inner Child 51

Chapter 9
- Know Where Your Pain Truly Comes From 55
- Label the Pain You Experience 56
- Keep A Journal of Your Feelings 57

Chapter 10
- Learn How to Identify Your Needs 60
- Create Rituals for Fulfilling Your Needs 62

Permit Yourself to Come First 63

Chapter 11
Learn How to Say "No" 66

Be Direct, Yet Polite When Asserting Yourself 67

Create A System of Personal Rules (And Follow Them) 68

Chapter 12
Learn to Forgive Your Mother in Your Way 73

Forgive Yourself for Your Experiences 74

Choose to Forgive Those Who Didn't Believe You 76

Chapter 13
Self-Validate Your Right to Seek Help 80

Find A Trauma-Informed Therapist 81

Create A Sense of Safety in Your Client-Therapist Relationship 81

Chapter 14
Reflect On Your Existing Relationships 84

Learn to Create Healthier Relationship Dynamics 85

Build A Healthier Social Circle 87

Chapter 15
Learn to Identify Red Flags and Respect Them 92

Reinforce Your Independence 93

Have A Plan for What You Will Do? 94

Conclusion 97

BOOK 2: SELF-LOVE WORKBOOK FOR WOMEN

Description 100

Introduction

    What Are Self-Care And Self-Love? 103

        Quick Advice 105

    Self-Love In Action – What It Looks And Feels Like 106

        Recognizing our Patterns 107

        Real-World Scenarios 107

    What Self-Love Isn't 109

        Takeaway 112

    Why Is Self-Love Important? 112

        Points to keep in mind 113

    A Journey Towards Self-Love In 10 Steps 114

Chapter 1

    What Are Self-Love Rituals And Why Are They So Powerful? .. 115

    Your "You" Time-Creating A Ritual Just For You 116

    Benefits Of Self-Love Rituals 122

Chapter 2

    Surrounding Yourself With Love And Support 125

    You At The Center 126

        How to build relationships? 127

    Communities Women Can Be Part Of 129

Chapter 3

    Journey Towards Self-Acceptance 131

The "What's Working for me" List 134
    Scheduling your 'What Works for Me' List takes discipline 136
Creating A Positive And Grateful Mindset.............................. 138
    The Origins of Positive Thinking in Modern Society 138
    The consolidation of positive thinking 139
    Positive thinking in the twenty-first century 139

Chapter 4

Being Mindful Of Our Body 145
Mindful Eating 148
    What are mindful eating habits? 150
Exercise To Look And Feel Good 151

Chapter 5

How Can Declutter Make Way For Self-Love? 157
Decluttering Towards More Freedom 158
    Get rid of that 'This stuff could be useful one-day Mentality' 159
Cleaning And Clearing Clutter 160
    The Method of Decluttering 160
    Decide to Take Action 161
    When in doubt, start with the floor 165
Valuing Space Over Stuff 166

Chapter 6

Understanding Self-Compassion And Its Benefits 174
The Three Elements Of Self-Compassion 175
    1. Self-kindness is not Self-Judgement 175

    2. Common humanity and not Isolation 176

    3. Mindfulness is not Over-Identification 176

    Benefits of Self-Compassion 177

Self-Compassion At Work 178

Self-Compassion In Relationships 180

Self-Compassion In Life 181

Dealing With Negativity 182

    Steps to deal with Negative thoughts and Events 186

    Learn to Forgive Yourself 188

Chapter 7

Learning To Meditate 193

    I'm Ready to Start Meditating, How Do I Begin? 194

Establishing A Practice 195

    Why Establish a Meditation Practice? 196

Benefits Of Self-Love Meditation 198

    Self-Love Meditation 199

    Your Past doesn't have to be your Future 200

Chapter 8

Boosting Your Self-Esteem By Focusing On What You Are Good At... 204

    List out your accomplishments 204

    Share Your Experience 205

    Use healthier motivation habits 205

    Take a 2-minute self-appreciation break 206

    Exploring Your Passions 206

Finding yourself- Is There Something You Already Love Doing? 207

Remember What You Loved as a Child 207

Eliminate Money from the Equation 208

Get feedback from Friends 209

Read through university or community college catalogs... 209

Identify your Heroes.. 209

Bottom line 210

Chapter 9

The Importance Of Having A Happy Place 212

Establishing A Happy Place 214

Chapter 10

Emotional Intelligence And Self-Love 217

Cultivating Emotional Intelligence... 218

Exercises to Cultivate Emotional Intelligence 220

Establishing your EQ Level 220

Identifying your Self-Awareness... 221

Identifying your Self-Management......................................223

Identifying your Social Awareness 224

Identifying your Relationship Management 225

Conclusion 229

NARCISSISTIC MOTHERS.

(IT'S NOT YOUR FAULT).

The Complete Guide to Understanding and Healing the Daughters of
Narcissistic Mothers, Healing Covert Emotional Abuse, Removing Guilt
Feelings and Finally Live Free

DESIRÉE SHANNON

Copyright © 2019 Desirée Shannon

All rights reserved

Introduction

Growing up with a narcissistic mother is a challenge unlike any other. When you are raised by a narcissistic mother, you experience unexplainable pressure from inside the relationship, outside of the relationship, and inside of yourself. This pressure commands you to be a certain person, a person who can shapeshift into any position that your narcissistic mother requires of you, even when she is not present to see it happen.

When I was growing up, the biggest challenges I faced were the fear of the abuse, and the growing realization that absolutely no one believed the abuse I was truly experiencing. At home, I was constantly walking on eggshells, trying not to trigger my mother and her abusive tendencies. Unfortunately, like many other narcissistic abuse sufferers, I had no idea what would trigger her and so I was constantly squirming and conforming to try to fit everchanging rules around who I was supposed to be and how I was supposed to behave. There was never any clarity around what the true expectations were, as they were constantly changing depending on her mood and who she needed me to be to fit her heinous standards on a day to day basis.

Outside of my home, I was still conforming to try and fit her expectations. I was terrified of behaving in a way that would displease her and her finding out, which would make my home life even worse. Or, worse, if I learned to stand up for myself and become my person when I was away from home and I brought that behavior home with me who knows what I would have faced. I was too scared to find out.

No matter how many times I reached out for help, no one seemed to believe how serious the abuse was. Everyone declared that it was just "typical mother and daughter relationship stuff." They believed I was a liar, a drama queen, a bad kid. They had no idea that they, too, had been groomed by my mother. Groomed to see me as a child that should be shunned and shamed, rather than a child who was desperately crying out for help and trying to be protected from the terrifying realities of my home life. She was an expert at keeping me in this position of never-ending abuse.

I swore when I was an adult I would get out of the situation and I would take control of myself. If I could just last a few more years, I'd be free. What I never realized was that by the time I was old enough to escape, I would be too deep in the mess. I wouldn't know how to escape, or where to go, to remove myself from my mother's abuse without losing everything and everyone else that mattered to me. I was trapped, and I had no idea what to do next. It was terrifying, gut-wrenching, and traumatizing.

Many years have passed since this time in my life, and so much has changed. I am no longer exposed to my mother's narcissism because I have learned to step away, create real boundaries, and protect myself. I have learned to heal the traumas I experienced and engage in healthy relationships with other people, and I have rebuilt my life in many ways. I still witness parts of myself that harbor the brokenness that I endured during those years of my life, but as they come up I have the power and the knowledge to heal them and continue moving forward in my life.

To this day, I still have a sense of longing in my heart for a relationship with my mother that could resemble what I truly needed, and what I truly need from her. However, I have come to accept that this is simply not possible and I have learned to fulfill these needs of being nurtured and loved from other areas of my life, including from within myself. While I don't think this feeling will ever fully go away, it no longer keeps me up at night feeling trapped in the worst possible version of a mental prison. I have learned to stand on my own two feet, and I am forever grateful for that.

In my healing journey, I have come to understand that narcissistic mothers are not nearly as uncommon as I once thought they were. Many daughters grow up with narcissistic mothers who expose them to similar abuse as I was exposed to as a child. I now realize that in healing, I also need to share my message and my knowledge to help other women heal from their traumatic mother-daughter relationships.

Take my story, and my advice, as being evidence that you can heal. There is hope after narcissistic abuse, and there is a way for you to feel whole and nourished in your life. That aching, longing, writhing feeling you hold onto the inside of you in every way is not a life sentence, and you are not doomed to experiencing abuse in all areas of your life for the rest of your life. You are more powerful than you know, and accessing your inner power starts with acknowledging the reality of your mother's wound and honoring and healing the pain that exists within it.

As you read through this book, I strongly encourage you to take your time and to surround yourself with loving people who can help you. Choose at least one or two people who believe you, and who can hold space for you as you uncover, face, and heal the traumas that you carry inside. Having this support will prove to be immensely powerful in helping you to stay grounded and focused during your healing experience. As well, take this at your own pace. There is no race in healing. This is your healing experience, and you need to do it in your way if it is truly going to support you with experiencing life changes.

Don't forget to leave a short review of this book on Amazon if you enjoy it, I'd love to hear your opinion!

If you are ready, let's begin.

Part One: What Is Narcissism?

CHAPTER 1

DEFINING NARCISSISTIC PERSONALITY DISORDER

One of the best tools I have used to heal myself and my traumas due to my relationship with my narcissistic mother is education. Educating yourself on everything that you can around narcissistic personality disorder, how it plays out in mother-daughter relationships, and how it has directly affected you as well as your symptoms can support you understand what you are experiencing. With this increased understanding, you can begin to identify where you are experiencing symptoms of this dysfunctional relationship in your life, why, and what can be done to help you heal them.

Let's start at the beginning of the equation: your mother. Your mother has likely been narcissistic since well before you were born, meaning you have been exposed to her narcissistic behaviors all your life. You have probably never known anything different. For this reason, we can conclude that the beginning of your problems around this starts with your mother and her disorder.

What Does Narcissistic Personality Disorder Look Like?

Narcissistic personality disorder itself is a mental condition that gives people an inflated sense of self-importance, dysfunctional relationships, an excessive deep-seated need for attention and admiration, and a complete lack of empathy for other people. Even though this is what they are experiencing, their symptoms can look somewhat different on the outside. This is because narcissists develop what are known as "masks" which are a sort of alter-ego that they hide behind to cover up the fact that there is something wrong with the way they behave. They use these masks to fake empathy, create false relationships with people in a way that does not reflect who they are truly are, and to create a genuine belief within themselves and others that there is nothing wrong with them.

The symptoms you are likely to see in a narcissist include an incredibly exaggerated sense of self-importance which often manifests as them behaving as if they are better than everyone else, and lying to make others believe it is true, too. They also have a sense of entitlement and feel as though they should get everything they want, including unlimited amounts of admiration for what they do. They often want to be recognized as superior to others, even without a reason to be recognized as so, and they fantasize about having unlimited success, power, brilliance, and beauty. A narcissist will also obsess over having the perfect everything in life, including the perfect house, the perfect mate, the perfect friends, and the perfect anything else. In their opinion, perfect things match their perfection, adding to their level of superiority over everyone else.

Narcissists can often be seen monopolizing conversations while also belittling or looking down on people who they believe to be inferior to themselves. They will often have a few people they seem to belittle the most, which are generally the ones who end up being their long-term victims for narcissistic abuse in many ways. However, they will belittle just about anyone as long as they can get away with it without tarnishing their perfect image, even though this image is all in their minds. Their inflated sense of self-importance also leads to them believing that they should get anything they want in life, no questions asked. They genuinely believe that people should just give in to what they desire without any struggles or difficulties in making it happen. Because of this, they will frequently take advantage of other people and treat them badly, with seemingly no understanding as to what they are doing and no compassion or empathy for the outcome of their actions.

Despite what it looks like on the outside, narcissists are highly envious of others which is largely the reason behind their constant bragging and self-inflating behaviors. They believe that when they act in these exaggerated ways that others envy them and wish to be them, which gives them an even larger feeling of self-importance.

The combination of all of their symptoms leads to narcissists being arrogant, haughty, conceited, pretentious and boastful. They are often the ones that appear confident to a fault, often to the point where people do not even believe that they are as confident as they truly claim to be. However, this is not always the case. Many narcissists have mastered experiencing an inflated sense of self-importance without coming across as overly confident, arrogant, or pretentious because they have come to realize that this does not serve their bigger mission. If they have come to realize that this behavior does not afford them the admiration and affection they desire, they will often tone down their behavior to increase the admiration and affection they receive. They can expertly shapeshift to fit any situation they need to, to get what they desire from that situation.

In private settings, narcissists are extremely unpredictable. They have severe interpersonal problems and can easily feel as though they have been wronged by others, leading to them reacting with rage. They may also react with rage or impatience if they feel that they are not receiving adequate special treatment from those around them. Narcissists struggle to deal with stress, regulate their emotions and behavior, and adapt to change. They can often become moody and depressed because they see their shortcomings and it makes them feel worse about themselves, which triggers their deep-seated insecurities, shame, and humiliation that they feel. These deep-seated feelings often come from realizing that they are not the same as others and fighting desperately to fit in, yet not knowing how to do so in a way that is not abusive and damaging to those around them. Even so, they will never admit this to anyone, under any circumstances.

WHAT CAUSES NARCISSISTIC PERSONALITY DISORDER?

The true cause or causes of narcissistic personality disorder are not known. Doctors cannot pinpoint any one thing that causes narcissism in people, so there is no way of knowing what may have caused your mother's condition. With that being said, some psychologists and doctors have come to suspect that a series of three things can contribute to narcissistic personality disorder. These three things include the environment, genetics, and neurobiology. There is no guarantee that these are the reasons, but they are suspected to at least contribute to the development of narcissistic personality disorder.

The environment is believed to affect the development of narcissism when an individual is raised by a parent who either excessively adores or excessively criticizes the child. If your mother was raised by parents who were poorly attuned to your mother's needs and who overly babied her or who excessively criticized her, this may have contributed to her narcissism. She may have also directly inherited it from her parents. As well, there may be an alteration in your mothers' brain that creates a disconnect between behaviors and thoughts, which contributes to the development of narcissistic personality disorder.

ARE THERE ANY CURES FOR NARCISSISTIC PERSONALITY DISORDER?

Because there are no known causes for narcissism, there is no known cure for it, either. Furthermore, most people who have narcissistic personality disorder will not acknowledge that they have it and so they will never take action to attempt to treat their narcissism. As a result, they end up having it for life and there are unlikely to be any improvements in their symptoms. Despite what you may think, if a person with narcissism does not want to admit that they have this disorder, which they are highly unlikely to, there is nothing you can do to make them see the truth. They will never truly see, accept, or own their behaviors and actions because this is in direct contradiction to what they are attempting to achieve.

If a narcissist does agree to get treatment, the best thing they can do is attend therapy. Independent therapy and family therapy are both methods that can be used to help a person with narcissism understand how they are affecting those around them, and how their behaviors can be changed and improved. In some cases, changes may occur and the individual may become easier to be around.

Chapter 2

The Narcissistic Mother

The way narcissism manifests in mothers specifically is unique, as children see their parents in a way that no one else does. Even a healthy relationship between a child and a parent is likely to be experienced in a way that is unique to them, not anyone else. Mothers tend to be more comfortable around their children, meaning that they can open up and express their true selves better around their children. For mothers who do not have narcissism, this is generally shown in a gentler and softer, nurturing manner. For those mothers who do have narcissism, however, this is generally a relationship where the narcissism will play out in a far more offensive and overwhelming manner than it would in any other relationship. In other words, narcissistic mothers tend to abuse their children, especially their daughters, more than anyone else.

Understanding how narcissism manifests in mothers is the best way to identify where your mother is abnormal compared to other mothers, and how these abnormalities are linked to her narcissistic personality disorder.

A Mother That is Threatened by Her Child

Narcissistic mothers often experience the feeling of being threatened by their children in the sense that they worry that their children are likely to take attention and admiration away from themselves. When narcissistic mothers notice their children are getting attention around any given subject, such as excelling in school, they will often begin to feel threatened and will attempt to minimize the value of the child's achievements.

A big way they do this is through how they talk to others, using sayings like:

- "Finally, you're good at something for once!"
- "It's about time you bring home an award for something."
- "Wait, you mean you did something good? Wow."

Speaking in a way that makes it seem like the child is otherwise terrible is a way that a narcissistic mother can control the amount of attention the child receives. They may receive attention around this one thing, but through her words, she tarnishes the child's reputation and therefore prevents the child from receiving further accolades anywhere else in their lives. This way, she can earn those accolades for herself and gain all of the excessive attention and admiration she needs from others.

An Effort at Self-Fulfillment Through You

Another big way that narcissistic mothers can be identified through their symptoms is through attempting to inflate their sense of self through you. My mother often did this by attempting to take credit for every positive thing I did in my life, making it appear as though she was the only reason, I had anything good going for myself. She would frequently use this as a way to take the attention away from me and put it on herself, even when it did not make reasonable sense for her to do so.

For example, as a child, I used to get excellent grades in school and every time my report card came home with several A's on it my mom would make dinner complete with all of my favorite things, which happened to be her favorites, too. She would go on to say how this was going to be an evening to celebrate me and my achievements, making me feel like maybe I had finally received her praise. Maybe she was finally proud of me. Every time, however, she would spend the entire dinner – my special dinner – talking about how she was responsible for my success. She would say things like how I would not be here without her, and how this proves that she is such a great mother, even going so far as to point out that I was cruel and mean for claiming otherwise when I called her out on her abuse to a family member one time

To add insult to injury, any time I would ask my mother for help with my homework she would either downright refuse or spend the entire time yelling at me for not being good enough, although it was her who was misunderstanding the assignments. In other words: I earned those A's in spite of her, not because of her. Over time I grew so resentful that I stopped caring about my grades at all because it was painful to have something, I was proud of ripped away and used as a tool against me constantly.

Narcissistic mothers frequently live through their children or use their children as a way to further inflate their sense of self. They generally do this because they know that at young ages children are not able to identify what is going on, and therefore they cannot stop the abuse from happening. By the time they are old enough to speak up, the mother has either made them too afraid to try or has already groomed everyone else to believe that the child is a problematic liar so that no one believes the child. In the end, the child is forced to live in a mental prison that is shaped and manned by their parent, which is a form of torture that no child should ever have to experience.

The Development of a Superficial Image

Another way that you can spot a narcissistic mother is in how they portray themselves to people outside of your family, or even outside of your relationship with her. Yes, narcissistic mothers will frequently wear several different masks even within one household, for example: abusing her child in private and pretending nothing ever happened when the child's father is around.

The development of a superficial image that portrays your mother as someone who never does anything wrong is a strategy that she uses to protect herself from her consequences. She does this to groom others into believing her, and not her child, which means that she can protect her primary source of being able to fulfill her cruel and unusual needs. This way, when she openly belittles and bullies her child the people around her believe it is warranted and the child has no hopes of escaping the experience.

Chapter 3

A Narcissistic Mother's Daughter

Symptoms in a narcissistic mother-daughter relationship are not exclusive to the mother. Daughters who have been raised by narcissistic mothers are also subjected to experiencing many of their painful symptoms that can lead to many problems in the future. You need to take the time to look at yourself as a part of the equation to see what symptoms you are experiencing, and to understand how they may be influencing you to experience more problems or abuse in your adult life.

Looking at your symptoms can be painful because you are going to have to face everything that you now experience and understand that this was all due to your mother. You might feel an intense amount of rage, sadness, pain, grief, guilt, or other emotions relating to these discoveries, so I strongly encourage you to make sure you can speak to someone you trust after reading this chapter. Being prepared to receive support as soon as you need it when difficult emotions or memories come up can be helpful in your recovery from this abuse.

You May Be Chronically Ashamed of Yourself

Daughters of narcissistic mothers are known for experiencing chronic shame in their lives, particularly around everything relating to who they are and what they do. It may feel like there is no limit to the shame that you experience, and that you tend to experience it in many different ways.

The shame that you experience now stems from always being made to feel inadequate as a child. Narcissistic mothers tend to be especially threatened by their daughters, which means that the level of abuse that you have experienced in terms of being put down and bullied is likely enormous. There is a good chance that your entire childhood was spent with you being told the many reasons as to why you were a bad person, and why you were not good enough. You were probably told that you were not deserving, not pretty, not smart, not worthy, and many other untrue things that were said to get you to stop bringing attention to yourself.

By making you feel horrible about yourself, your mother could feel confident that you would stay quiet and hidden all on your own so that she did not have to attempt to do it for you. She also would not have to take responsibility for dragging your name through the mud or spreading bad rumors about you, which is a common narcissistic behavior known as "smearing." In some ways, your mom may have even used your low self-esteem to increase her sense of importance, such as by bragging about how she had to stand up for you or try to build you up in certain situations because you lack self-esteem. Of course, she would never mention that your lack of self-esteem came from her in the first place because this would take away from her perfect image.

As an adult, you may now experience chronic shame around everything in your life even when you know it is not needed. You might hold yourself to unreasonably high standards, feel guilty about things that are normal human experiences, and attempt to behave like a superhuman because you have been told that you are not good enough. These behaviors are likely both an effort to be seen as a good person and an effort to avoid being abused any further because in your childhood you would be abused if you did not fight to achieve these unreasonable standards. This shame is extremely toxic, painful, and life-altering, which is why we are going to spend so much time addressing and healing it in part 2 of this book.

CHILDHOOD ABUSE MAY LEAD TO ADULTHOOD ABUSE OR TOXIC RELATIONSHIP PATTERNS

Any form of childhood abuse can lead to children growing up and entering abusive and toxic relationships, and a child abused by a narcissist is no different. It is possible that as an adult you are now finding yourself in many toxic relationships, or relationships that are even downright abusive. You might feel like you have some sort of hidden "signal" that somehow calls in people who will take advantage of you, bully you, or abuse you through narcissism in your adult life. Many daughters of narcissistic mothers feel as though they cannot get away from narcissism, even though they were sure that leaving their childhood homes would suffice.

The reason why you may be experiencing toxic or abusive relationships now in adulthood is that you have never been taught boundaries or important self-care steps in life. Being raised by someone who commanded you to live your entire life based on her needs and desires has resulted in you not knowing how to fully stand up for yourself and take care of yourself in relationships now. This may be painful to admit, but, indeed, it is likely the reason why this is happening. If you notice that you seem to be surrounded by people who abuse you or take advantage of you and you cannot seem to understand why this happens, there is a good chance that it is a product of your groomed behaviors.

YOU MAY REFLECT SOME OF YOUR MOTHER'S SYMPTOMS

As a daughter of a narcissist, this can be one of the more scary symptoms that you may face. It is one thing to feel unsafe with others, but to feel unsafe within yourself and to recognize yourself behaving in ways that you do not like can be downright terrifying. There is a chance that now as an adult you reflect some of your mother's symptoms, and this can lead to an intense fear that you are going to become abusive toward someone you love just like she did. You may not understand why these behaviors or exist or have any clarity around how far they will develop too, leaving you feeling powerless and as though it may be inevitable for you to follow in her damaging footsteps.

Believe it or not, even though many people do not like to talk about this point, it is quite common in those who survive narcissistic abuse from a parent specifically. The reason for this symptom is that as a child you are supposed to be raised by a nurturing guardian who can guide you to learn how to navigate various parts of life. Ideally, a healthy guardian should have taught you how to deal with difficult emotions, conflict, expectations, self-esteem, insecurities, and other natural parts of life. Unfortunately, you were raised by a mom who did not know how and who regularly modeled extremely poor examples of how an individual should deal with these things. As an adult, you reflecting this behavior is unlikely to be you displaying true narcissism and more likely to be you displaying poor coping methods in life. With proper healing and efforts, you should be able to identify new ways for you to cope with things in life, enabling you to move beyond the patterns of repeating your mothers' behavior due to not knowing a better way.

THERE MAY BE THE FEELING OF A DEEP VOID IN YOUR LIFE

One of the most painful things that I have experienced as a daughter of a narcissist, even to this day, is that void that you feel in yourself and your life around your mother. As an adult, you may now find yourself longing for a positive relationship with your mother, possibly to the point where you keep attempting to have a better relationship with her only to find yourself trapped in the cycle again and again. This is a common experience for daughters of narcissistic mothers and I want to tell you right now that this is not a poor reflection of you, instead, it is a painful reflection of your reality.

Even when you heal yourself from your mother's abuse, you are likely to find yourself in moments where you wish you had a healthy, supportive mother to rely on. You might even recall the times your mother showed you her charming mask, leading you to feel like maybe you can call her for support on just this one thing, hoping that she will offer that type of charm and support once again. It can be painful when you realize that your mother is unavailable to offer you the support and the love that you need, and even more painful when you realize that she has no idea why you feel so disconnected and alone in the world due to her treatment. This is a natural part of the recovery and healing process, and in time it does become a lot easier to navigate. While the pain itself is always there, you will find that you become much stronger in healing that pain and coping with it when it rears its head. This way, you do not put yourself in a game of yo-yo with trying to get your mother to be the woman you need her to be when she truly can't be.

Chapter 4

The Future of Your Relationship

The future of what your relationship will look like with your mother is ultimately going to depend on you and what you think will be the best for your situation. With that being said, I strongly advise taking a lengthy break from talking to your mom while you heal yourself from her abuse and then ease yourself back into any sort of relationship you might share if this is the path you choose. Attempting to heal from your mom's abuse while keeping yourself trapped in the cycle by maintaining a fairly close relationship, or at least a consistent relationship, during the healing cycle can disrupt your results. You might find yourself constantly getting dragged back in despite how much effort you put into healing, which can leave you feeling extremely poorly about yourself.

With narcissistic mothers there are generally three ways that the relationship can go: you can break away entirely, you can have a small relationship, or you can have a consistent relationship with strong boundaries. What you choose will depend on your chosen coping methods and the level of relationship that you can personally handle without feeling impacted by her abuse. This means that after your break you should slowly build your relationship back up and not exceed what feels right for you, to ensure you do not get sucked into old behaviors that could lead to a complete relapse in your relationship.

What to Do If Your Relationship Must End Completely

The idea that your relationship with your mother might need to end completely can be incredibly painful, especially if you have spent a large portion of your life hoping it would get better. Until this point in your life, you may have been under the influence of the belief that you could somehow contort yourself to make things better and that this would lead to your mom like you more and your relationship is fixed. Unfortunately, this is not real and there is no true hope of your relationship ever being the one that you want it to be, as hard as that is to admit. Believe me, it took me a long time and many relapses in my relationship with my mother to realize that she was never going to be the nurturing, supportive, loving mother that I wanted and needed.

If you find yourself in a position where your relationship must end completely, it might be since your mother's abuse is extreme, possibly on the brink of violent, or causing severe toxicity and trauma in your life. Your mother may be abusive to the point where you cannot have even one conversation with her without her creating a web of abuse, which leads to you feeling like you need to end the relationship completely. In this case, what you need to do is completely cut all ties and keep those ties cut. If you find yourself in a situation where the severity of the narcissism is so advanced that you must cut ties, you must remember why the situation got this advanced. When you find yourself wanting to relapse into a relationship with your mother, you must remember the reason why you no longer have a relationship with her in the first place. If you go back and forth in relationships that are this damaging it can be even more damaging as you begin to experience the trauma from your mother, as well as the trauma from yourself each time you "allow" yourself to get sucked in. This can become a huge point of guilt, and it can make healing even harder, so it is strongly advised that if you make this decision you stick to it.

What to Do If You Need to Minimize Your Relationship

In some situations, you may not need to, or maybe you can't, completely end your relationship with your mother. In this case, it is ideal that you minimize your relationship with her. Minimizing your relationship can look however you want it to look, but ultimately it requires you to avoid seeing or talking to your mother consistently. You might find yourself only talking to her when it's the holidays and you are together at a family gathering, or possibly up to once or twice a month. The frequency of this relationship ultimately depends on you and what you genuinely feel that you can handle with your mother.

This is the area where I fall with my mother. The rest of my family is quite close and I want to make sure that I maintain a relationship with them, which inevitably means that I need to be around my mother from time to time. Aside from these visits, however, I do not contact my mother because it does not feel right for me to do so. I feel stronger when I experience life on my own than I do when I attempt to celebrate with or confide in my mother only to be met with emotional unavailability and abuse. For that reason, this is my best coping method. Even with the minimal amounts of time we see and talk to each other, it still takes immense strength for me to stand strong in my coping methods and refrain from getting sucked into my mother's drama and abuse.

WHAT TO DO IF YOU NEED TO STAY CONSISTENT IN YOUR RELATIONSHIP

Some daughters will continue to have a fairly consistent relationship with their mother, even after they heal from narcissism. This is often very uncommon, however, as it can be extremely challenging to remain truly removed from the dysfunction when you are still regularly being exposed to your mother and all of her symptoms. The daughters who do find themselves capable of consistently communicating with their mothers and maintaining high-frequency relationships require massive amounts of strength to be able to uphold their boundaries and stay strong. It is incredibly challenging to break the dynamic between the mother and daughter in this scenario because the mother already has it so ingrained in her, and it is all the daughter has known since birth. In these relationships, the mother often knows exactly what to say to push the buttons of her daughter to force her back into the abuse cycle.

Due to the complexity of narcissism and the tact and calculated abuse they dish out, it is important to realize that the likelihood of you being able to maintain a consistent relationship with your mother and heal from her abuse is highly unlikely. If you do attempt to retain this type of relationship, there is a good chance that you are doing so because of her grooming and conditioning to force you to believe that it is required and that you are somehow a bad person if you don't. It may even be due to her smearing you and abusing you if you do try to stand up for yourself and get away from the abuse.

Make sure that if you are going to try this that you strongly consider why you are doing it and that if you must, you constantly work on increasing your strength and boundaries and upholding them in your relationship. You can never let your guard down here, or your mother will see the opportunity and attempt to take advantage of it. No matter how far you may get with protecting yourself, your mother will always be attempting to abuse you throughout your entire life. She will likely even go so far as to use compliance as a way to show you that the relationship can be "all better" to reel you in, just to start the dynamic all over again. You must always be cautious and in control of this relationship, no matter what. For that reason, it is likely going to be far too draining for you to uphold and it is not a good idea to aim for this type of relationship.

Chapter 5

Beliefs You Have Developed

The final element of the entire puzzle of your trauma from your mother's narcissism is your belief system. Your belief system is a system of beliefs that you cultivate throughout your childhood, adolescent, and adult life that help dictate how you think, perceive and feel about the world around you. Psychologists say that until you are 7 years old, the beliefs you have are entirely taken on by your parents and other authority influences in your life and that any beliefs you create in this time of your life lay the foundation for the rest of your experience. While you can change beliefs from this period of your life, you can only do so if you can become aware of those beliefs and do the healing work on them.

Identifying your beliefs now is going to help you have a strong understanding of the entire image of your experiences with your mother, which will enable you to create the strongest healing plan for yourself. The concept behind this is that the more you are aware of your wounds and why they exist, the more you can heal.

Some of the beliefs that you may have picked up from your childhood include ones like you live only to serve others, and you are not good enough or worthy enough of having a good life. There are many other beliefs that you may have picked up, however, that you are going to need to consider so that you can start healing them. Some of them you might find in this chapter, others you might find as you go through your healing journey and become aware of them through the healing practices you will engage in. There is no right or wrong answer as to what beliefs could have formed from this relationship, so take note of all of them. Use this chapter, however, to start laying the foundation of this understanding.

You Live Only to Serve Others

This is one of the biggest beliefs that daughters of narcissistic mothers develop in their lives. The belief that you only live to serve others can manifest as people-pleasing behaviors and codependency, so if you find yourself behaving in either of these patterns it may be due to an underlying belief that you are only here to serve others. This belief is formed when your mother makes it clear to you from a young age that you must please her and serve her every need for her to like you. Otherwise, she is likely to withhold affection, respect, and love from you throughout your life. As you see this playing out, you develop a deep-seated belief that your role in the world is to serve others and that serving yourself in any way, including through taking basic care of yourself, is selfish and bad. Alongside this belief, you might also find yourself attempting to shrink yourself and minimize the amount of space you take up in a room both physically and in terms of your actual presence to avoid being seen. In your belief system, being seen may feel like you are taking attention away from someone else which you have been taught is a bad thing due to your mother's need for excessive amounts of admiration and attention.

You Cannot Do Anything Right

As you grow up with a narcissistic mother there is always volatility to the situation because you can never tell what is going to set her off or cause her to punish you in some way, either through withholding nurturing or through rage. Through her behavior, you begin to realize that she could blow up at anything you do, which can lead to you feeling like you can never do anything right. It is likely that through her words she also reinforces the belief that you are bad at everything, making you truly believe that you are incapable.

Your Perception of Reality is Not True

A big and often dangerous belief system that daughters of narcissistic mothers grow up with is that their perception of reality is not true. This mistrust in their perception of reality leads to the belief that they are incapable of recalling events as they happened, and it is caused by narcissistic mothers lying to others and to their daughters to cover up their abuse. This chronic lying and twisting of events can make it feel like you are incapable of remembering anything factually, which leaves you believing that you have to rely on others to recount facts for you because you are incapable. This can be particularly dangerous as it can completely distort your sense of reality and leave you trusting in the wrong people to provide you with a true sense of reality. The truth of the matter is, the only positive way to work through this belief system is to reinforce your trust in yourself and learn to believe in your perception of reality. It is never a good or safe idea to place this trust and task outside of yourself as it will leave you vulnerable to abuse while also keeping your sense of self-esteem, self-confidence, and self-trust low.

PART TWO:

HEALING AND FREEING YOURSELF

Chapter 6

Prepare Yourself for the Experience

Now that you are fully aware of what the entire spectrum of your abuse dynamic looks like, or at least aware of the basic foundation of it all, it is time for you to begin the process of healing from the abuse that you have endured. Before you can dive into the healing process, I strongly advise that you prepare yourself for what the healing process is going to be like.

For many individuals healing from narcissistic abuse and trauma, knowing what to expect can make navigating the healing process a lot easier. Without the ability to navigate it clearly, you can find yourself feeling almost a sense of shock for what you are facing which can leave you wanting to return to the abusive situation to regain comfort. It may sound silly from the outside, but from the perspective of someone who has been abused, it makes sense that they would desperately cling to any sensation of comfort, even if that comfort is coming from a toxic place. When you live an incredibly uncomfortable life, you will take whatever scraps you can get to feel better and losing touch with those can feel vulnerable and scary, especially when you know how much danger you could be exposed to in doing so.

When I was recovering from my narcissistic mother's abuse, I went back and forth a few times before finally escaping the dynamic. It took me some time to completely accept what I was going through and the pain that I was experiencing and to be able to endure the discomfort I felt during the transition from being the codependent daughter of a narcissist to being an independent adult. Preparing yourself may help you make the move more effectively and transition in a way that is less traumatic and overwhelming to you as you seek to live a healthier life for yourself.

There are multiple steps you can take to help you prepare to move out of the traumatic dynamic and into one that is going to be more stable and healthy for you. You must complete all of them before you start making any bigger moves, as this will give you something stable to transition with, which will make it easier for you to completely escape from the dynamic.

Understand the Cycle of Narcissism

To truly understand what you are about to go through when you break yourself out of the cycle of narcissism, you need to understand what the cycle of narcissism is and what role you play in this abuse system. The narcissistic cycle of abuse has four basic steps in it that enables narcissists to gain the admiration and attention they desire while keeping their sense of superiority over others. This calculated cycle helps them smear your name and keep you quiet while enabling them to keep going without ever getting caught.

The narcissistic abuse cycle starts with the narcissist feeling threatened, which causes them to abuse others while also positioning themselves as the victim. Through this, they feel empowered which enables them to feel superior before they begin feeling threatened all over again. What will threaten a narcissist varies from person to person, and can also vary from day to day. For example, sometimes, a narcissistic mother might feel threatened by their daughter looking prettier than they do because it gets the daughter more attention, whereas other times she might be fine with it because it makes her look good. There is never a way to truly tell what position she is going to take, which makes the narcissist mother volatile and frustrating to be around.

For the victim of the abuse or the daughter in this situation, it can feel like your mother builds you up and makes you feel great at times, only to tear you down again. After tearing you down she leaves you all but begging her for more positive attention, only to withhold it from you as a form of punishment. Then, when you finally give up and you have no energy left to fight, she throws you a small amount of attention to give you a reason to start fighting for her attention and admiration all over again. This cycle continues as you constantly do whatever you can just to receive those small scraps of positive attention from your mother, despite never knowing when they are coming or what might warrant them.

Surround Yourself with Supportive Outsiders

For you to truly prepare to break free from the aforementioned cycle, you need to make sure that you have surrounded yourself with a support team that can help you get through the challenging periods that you are about to face. You want to make sure that your support team is filled with people that you can trust, and that will help you feel safe and comforted during the process. They should be people that you can inform as to what is going on and what you plan on doing, and that will believe your experiences and support you in the process.

There are two types of people you should surround yourself with when it comes to leaving an abusive situation of any kind: friends and loved ones, and trained professionals. Both are going to help you navigate what you are about to face more effectively so that you can withstand the transition period that takes place between you breaking out of the cycle and feeling confident and comfortable where you stand after the fact.

Loved ones can be the more challenging part of your support system to create, as you truly need to make sure that the loved ones you rely on are going to be helpful and will support you in leaving the toxic situation. If you surround yourself with loved ones who are likely to side with your mother, or who will try to downplay the reality of your experiences, you might find yourself falling back into the cycle all over again because you are trusting someone who is unhelpful. If you have people in your social circle who will genuinely believe you and who will not attempt to pressure you into healing your relationship with your mom, rather than healing yourself from her abuse, choose these people to surround yourself with. Otherwise, focus on spending some time intentionally making new friends who you can surround yourself with. This may take some time, but if you do so intentionally you will find that you can create at least two or three strong connections with people who will serve as your support team through the transition period. Make sure these are friends you can keep long term, as you do not want to use anyone, nor do you want to attempt to create a false sense of connection just so you can leave an unhealthy relationship.

The people you surround yourself with who are considered loved ones, such as friends and family, are the ones who you are going to talk to about your experiences. It is a good idea to open up to them honestly about what you are going through and what you may face in the future so that they understand what you are experiencing. This way, if you do need someone to talk to or vent to during the process they do not feel blindsided by your request to talk about something heavy. As well, they can look out for you and help you identify possible patterns that may lead to you going back into the unhealthy relationship with your mother.

You also need to have professionals on your team, as your loved ones are simply not going to be able to help with everything. Furthermore, attempting to rely on them for that emotional support too often can lead to you feeling codependent on them which can transition your problems into new relationships. Having a therapist on your support team can give you someone qualified to talk to about the deeper and heavier experiences that can also guide you to begin healing from your abuse. This is a great way to ensure that you are taking responsible action toward managing the pain and troubles you are facing due to your mom, and it is a good idea to do so as soon as you plan on healing from the relationship you share with your mother. We talk in great depth about how you can find a therapist and work with one in chapter 13, as I know the idea of working with a therapist can be terrifying for someone who is going through the process of healing from this type of abuse.

Prepare Yourself For The Emotional Experience

The final straw in preparing yourself for breaking free from the cycle of narcissistic abuse is preparing yourself for the emotional experience you are going to have throughout this process. Healing from any form of abuse is painful, sometimes to the point where it feels even more painful than enduring the abuse itself. This is because when you are within the abuse you enter "survival mode" which ultimately keeps you fairly numb to the greater picture of what is going on. While you certainly experience emotions, you do not experience them as deeply as you do when you finally start to come out of it. Furthermore, stepping out of the cycle is going to expose you to emotions you are afraid of: such as a fear of being punished for not doing what are "supposed to do" or the anxiety of standing on your own two feet when you are used to serving someone else. When you were a child, these experiences lead to massive punishment and experiences of neglect and abuse. As an adult, your mother may attempt to engage in these forms of punishment again, but the reality is that you are no longer at the mercy of her. You are no longer a young child living in her house and relying on her to feed you, clothe you, house you, and take care of your needs. As scary as it may feel now, you are capable of standing on your own two feet and building your own life away from your mother's abuse and neglect.

The best way to prepare yourself for the emotional aspect of breaking free is to read stories about other daughters who have done it. You may even be able to find an online support forum of other daughters' who have experienced it so that you can communicate with people who truly understand what it feels like to go through what you have gone through. Listening to other people's stories and hearing what they truly experienced will help you mentally brace yourself for what is coming so that the intense emotional experience does not come as a shock and create a traumatic relapse in your relationship.

Chapter 7

Getting Perspective

Once you have your foundation in place for you to begin taking steps toward breaking free from the abuse pattern, you need to focus on getting perspective on what you are going through. For me, perspective was the difference between blaming myself and feeling completely consumed by what I was going through and realizing that it was not my fault and there was a world beyond the abusive one my mother dragged me into.

When you have been exposed to narcissistic abuse from your mother since a young age, your entire perspective is shaped around what she pressured you to think. For this reason, it can be easy to get sucked into the anxiety and fear that comes with this chaotic perspective that never seems to have any clear sense of direction or rules around what it truly is, or isn't. The perspective of narcissists and therefore the perspective of their abusers is confusing, overwhelming, and always filled with uncertainty which can further increase your feelings of stress related to the situation. This chaotic energy becomes so consuming that you cannot see anything beyond it, both because you were groomed not to and because you are constantly just trying to figure things out. It's like being trapped in a puzzle and not knowing the way out, and never realizing that you now have the power to put the puzzle down and walk away from it entirely.

Gaining perspective helps you see your situation objectively and recognize each part of the situation for what it truly is. In many ways, it cultivates a deeper sense of awareness around the chaotic energy you have been experiencing and why it has been so all-consuming for all of these years. You will find that in doing this, you start to put the pieces of reality together and the concept that there is a reality beyond this confusing and painful web no longer seems so crazy.

Start Viewing Your Situation for What it Truly Is

The first thing you need to do to start gaining perspective in your experience is to start viewing the situation for what it truly is. Begin to see your relationship with your mother as being one that is dysfunctional and filled with abuse.

Pull yourself out of the perspective of being a daughter who is never good enough, or who constantly has to try to live up to your mother's unreasonable and unclear standards of who you should be. Stop trying to understand it from your mother's perspective, because it is never going to make sense: it is always going to shift just as soon as you think you have it figured out so that she can keep you on your toes and pleasing her as much as you can. Let go of the idea that this reality she has attempted to spin for you is one that you are required to live in, and stop trying to justify why she behaves the way she behaves.

Instead: begin to label everything accordingly. Identify yourself as being the victim of narcissistic abuse, and your mother as being a narcissist. Identify the trauma you experienced in your childhood, and the coping methods you used to deal with that trauma, as being the product of abuse, not the product of you being a bad kid.

If you can, begin to label the parts of the narcissistic abuse cycle in your relationship and develop an understanding as to how it plays out specifically in your relationship. The cycle itself is virtually always the same from relationship to relationship, but what specific topics or techniques your mother may use to get you to comply with her may vary. For example, my mother always used the topic of my beauty and my artistic talent to either make me feel inadequate or to make herself look better. Your mother may have used your fitness level, your knowledge, or any other talents or skills you may have against you or in favor of herself throughout your childhood and even now into your adult years.

When you can begin to identify these patterns, especially when you can start labeling them while they are in action as you actively experience them, it becomes easier for you to see them for what they are. This way, you can start distancing yourself from the distorted reality of your mother and start centering yourself in the truth, your truth.

Explain the Story from An Outsider's Perspective

A great way to begin practically and honestly building your perspective around your mother, your trauma, and the acts of abuse you have experienced is to explain the story from an outsider's perspective. This practice was particularly emotional for me as it forced me to face the reality of just how much abuse I had faced in my life and how numb I had become to most of it. Sure, it hurt, but I had become so used to it that it no longer hurt as bad as it should, and that piece reality itself was extremely painful.

To explain the story from an outsider's perspective, sit down with a journal and begin to write down your entire life story surrounding your mother. Write about the types of abuse you faced, the stories she told you, and the ways she treated you when it was just the two of you versus when other people were around. Also be sure to write down what you overheard her telling other people about you, and what she seemed to behave like when she believed you were not around. Get a full perspective around what was going on and write it from third person so that it feels like you are truly narrating your life story with your mother.

You might find that using this outside perspective helps you fully see and become aware of everything that was going on. Doing this, starting as early as your youngest memories, can help you put into perspective the abuse you have faced and the way it has affected you throughout your lifetime.

In doing this, I also encourage you to start thinking about how you might respond to someone else telling you these experiences. Treat yourself with the same level of respect, concern, compassion, and support that you would treat someone else who was going through what you have gone through. Learning to give yourself this level of care will help you begin to honor what you have truly come through and accepted the level of trauma that you have faced in your life. Truly accepting and admitting to those levels of trauma and the way they have affected you will help you begin to have compassion for yourself and develop a sense of self-awareness around your trauma which will prove to be extremely helpful in healing from it.

Keep the Reality in Mind When Making Your Exit

When you begin to take steps toward breaking free from your abuse cycle, you must keep this reality in mind. Keep your journal nearby and re-read it anytime you feel yourself getting pulled into your mother's distorted reality so that you can keep yourself focused on your recovery. The more you can reinforce your sense of reality, the less likely you will be to fall for her abuse and find yourself believing in her lies again. This is your best opportunity to free yourself from abuse and move forward without constant relapses and regressions into the cycle.

Chapter 8

Validating Yourself

As you begin to build your perspective around what you have gone through, you may find that you have a hard time believing in your perspective. Until now, your mother has taught you to believe in her and everything she tells you, even if you can compile a long list of evidence as to why she is wrong and why her sense of reality is distorted. Because of the way your mother has groomed you, you will find yourself believing in what she says even if what she says makes no sense.

Another thing you might feel when you start to develop your perspective is intense fear. I knew that in the past when I had my perspective and trusted in myself and my non-distorted sense of reality it leads to a lot of abuse when I was younger, which created a large amount of anxiety in my perspective as an adult. I had severe anxiety around believing in myself and believing in my sense of reality because I had always been told that I was wrong and if I chose to believe I wasn't then I was severely punished. Maybe the same happened for you, or something along those similar lines.

In addition to disproving your sense of reality, your mother also likely destroyed your ability to validate yourself in many other ways. Through forcing you to fight for her admiration and attention she may have left you feeling inadequate in several different ways in your life. She may have left you feeling inadequate in every way. How deep your traumas lie and where they lie will depend on your unique relationship with your mom, but a common experience between those who have been abused by narcissists is to have major struggles invalidating themselves. Regardless of what you struggle to validate yourself around, I can almost guarantee you experience this symptom.

You must begin to learn to validate yourself before you start the process of officially physically breaking free from the cycle of abuse with your mother so that as you go through the transition you can validate yourself. If you do not practice validating yourself first, you will find that as much as you want to break free you struggle to do so because you do not have her validation to help you do it. Even though this may sound illogical, the process of breaking free will distort your sense of logic as you will be deeply invested in your emotions which can lead you to defy logic altogether. This is exactly how abuse gets so far in the first place.

Start to Get to Know Yourself in A Deeper Way

One of the biggest ways that you can begin to validate yourself is through getting to know yourself more deeply. At this point, your knowledge in who you truly are maybe extremely limited as you have been fed your sense of identity by your mother and you have not had time to find out whether it was true or not due to her excessive demands. Spending time getting to know yourself helps you feel a greater sense of confidence in who you are so that you can validate yourself as you go through life. When you know who you are, it becomes easier to assert yourself to both yourself and others, effectively giving you a strong foundation for self-validation.

After an abusive relationship, you should always spend time getting to know yourself all over again, even if you think you already know yourself. Abusive relationships can majorly distort your sense of identity, especially when they start in childhood, and you do not want to find yourself holding onto a distorted sense of identity automatically because you never dug deeper. Focus on starting with the very basics, such as who you are, where you work, and what you like and dislike. From there, spend time getting to know yourself more intimately by behaving as if you were getting to know a brand new friend. Through affording yourself this level of attention to detail, you can give yourself the best opportunity to get to know who you truly are and begin validating that true version of yourself to yourself and others in your life.

Keep Your Promises to Yourself, Always

Part of being able to validate yourself is being able to trust yourself. Many of the different abusive strategies your mother has used against you in your life will have directly compromised your ability to trust in yourself, so you need to make this a priority. A big way that you can start creating a deeper sense of trust in yourself right now is through keeping your promises to yourself, always, such as the promises you make to yourself throughout this book. When you promise that you are going to get to know yourself better, for example, keep that promise and put the effort in.

Keep promises to yourself no matter how big or small they are because this starts to give you a sense of self-sufficiency. When you can keep promises to yourself, you realize that you are capable of doing things that are important to you which means that you are independent and capable of standing on your own two feet. This works the same as achieving goals does the more goals you achieve, the more you can achieve because the more you believe in yourself. Teach yourself that you are trustworthy and capable of making promises to yourself and keeping every single one.

Learn to Tell Yourself the Things You Need to Hear

When we seek validation from outside of ourselves, we are directly looking for someone to tell us something that we believe we need to hear from another individual. For example, if you start asking other people to tell you whether or not your work is well done because you want to hear that it is, you believe that your work is not good enough until someone else decides that this is true.

Constantly seeking validation from outside of yourself in the form of approval is a direct reflection of how narcissists abuse their victims, so understand that this is something you were taught to do from a very young age. Still, as an adult, it is your responsibility to change this behavior if you are going to be able to give yourself the best foundation to leap from to remove yourself from your mother's abuse.

When you find yourself searching for validation from outside of yourself, ask yourself what it is that you truly want to hear from someone else. What message are you looking for that you believe you need to receive from someone else for it to be true? Once you have discovered what that message is, begin affirming it to yourself as much as you need to for you to believe it to be true. This is a great step for actually, actively validating yourself in every area of your life.

Stop Judging Yourself and Your Feelings

When you struggle to validate yourself, one thing you likely notice is that you judge yourself and your feelings, likely in the same ways that your mother has judged you. This judgment comes from outside of yourself, much in the same way that validation has always come from outside of yourself. In a sense, this is you using external validation to validate why you are not good enough at something, which is essentially the same thing as looking outside of yourself for validation that you are good enough.

To start validating yourself more effectively, you need to witness this judgment and stop using it as a way to punish yourself or put yourself down by validating why you are bad. Remember that this is validation coming from outside of yourself and that you do not need validation from outside of yourself because you can validate yourself. Then, begin to identify what you truly believe about what you were originally judging, such as a behavior or a feeling, and start affirming that to yourself. For example, if your mother judges you for wearing your hair down and you think you look pretty wearing it down, start wearing it down and validating to yourself that you like it and therefore it is pretty.

PRACTICE PARENTING YOUR INNER CHILD

A massive action step that you can begin taking toward healing yourself from your mother's narcissistic abuse is learning how to parent your inner child. In psychology, our inner child is that inner part of ourselves that still exists as an imprint from our childhood. This is the inner part of yourself that might still be afraid of the dark even if you behave bravely now, or that might believe in the boogie man even though as an adult you know he's not real. Despite how silly it might sound, this part of yourself is very real and continues to exist inside of yourself even though you are now a full-grown adult living your own life.

When you are abused, such as through narcissism, your inner child exists with a lot more than just a fear of the dark or a made-up monster that hides under your bed. Your inner child exists with fear around a very real abuser who may or may not still be a part of your adult life. If you have already broken ties with her, she may still have a grasp on the way you think and feel, or on the way other family members perceive you, even all of these years later, causing you to indirectly experience her abuse. In all of these scenarios, she is still affecting you which means your inner child is still terrified of being exposed to her abuse.

A great way to identify who your inner child is would be to recognize the difference between how you talk to a trusted person about your mother, versus how you act around your mother. For example, with my fiancé, I used to pinpoint exactly how my mother was abusive and what I should do to get away from it, but the minute she was around I would feel speechless and powerless. This was the difference between my adult self and my inner child.

Once you have identified your inner child, and you begin to see where the lines exist between who she is and who you are now, you can start to parent your inner child. This means you can talk to that part of yourself as if she is with you right now and you are parenting her in the ways that your mom never did. You can tell yourself "it's okay, I'm here now and I'm going to protect you" or "she can't hurt you anymore, I'm here to take care of you." Learning how to talk to your inner child in this way builds up your belief in yourself and helps this fearful part of yourself feel seen, protected, and validated by you. As a result, not only do you increase your sense of self-validation but you also create an opportunity for you to boost your confidence and your sense of self-esteem, making it even easier for you to grow and heal from your narcissistic mother.

Chapter 9

Acknowledging Your Pain

The more you dig into the realization of what has happened to you and what you have gone through, and stop blocking it out and justifying the behaviors of your mother, the more pain you are likely to experience. Coming into this process, you were already experiencing the pain of your mother's abuse, but it may start to feel like it is stinging even more as you begin to fully acknowledge it and accept what has happened to you over the years. It may start to feel like you are unraveling many pieces of your story that you kept blocked out or hidden for fear of having to face the growing amounts of pain that were accumulating in your life. As you do, this pain can all come up at once and it can be incredibly challenging. What's worse is you may be tempted to numb it out again or justify it in favor of your mother and her behaviors because this is how she has groomed you to act in order to continue serving her abusive patterns, by keeping you the victim.

I will repeat what I have already said a few times here, but this is where it is really important that you become aware of what you are experiencing and you consciously work through it. This is the point where I would always become overwhelmed by emotion and I would backtrack into a relapsed relationship with my mother because I was too afraid to go through the pain of being in this state. The problem with each backtracks was that each time I realized more about the truth of what I was living and it became harder to numb everything out and stay trapped in the abuse. As I continued to realize how toxic and dangerous these behaviors were, I found myself feeling even more traumatized by each succeeding episode of abuse. At this point, not only was I being traumatized by my mothers' abuse but I was also being traumatized by my seeming inability to get beyond it.

I want to help you avoid this repetitive cycle I found myself in and fully release yourself from this abuse pattern so that you can avoid the intense trauma that comes with the back-and-forth of trying to escape. The better you can prepare yourself for this and work through it, the better your chances are going to be for you to fully remove yourself from the situation and truly begin healing yourself and your life. Understand that this does not mean you are not going to experience trauma or the painful reality of your previous trauma, but it will hopefully help you avoid compounding your trauma with additional traumatizing experiences.

At this point in your journey, if you have not already you should begin distancing yourself from your mother. While you go through this painful acknowledgment you do not want to have your mother exercising her narcissistic tools to attempt to pull you back in, leaving you feeling even more confused and overwhelmed. You truly want to be as removed from the situation as possible so that you can rely on yourself and your sense of reality to help guide you through the healing process. As you go through this distancing there will be even more pain relating to the distancing itself that comes up, so be prepared to be patient with yourself and move at a pace that fits what you can reasonably handle. It is perfectly safe to take this all at your own pace, as attempting to rush any step can lead to you feeling overwhelmed and regressing into the victim mindset which will keep you trapped in the cycle even longer.

Know Where Your Pain Truly Comes From

The first thing you need to do when you are acknowledging the pain you are experiencing is taking the time to understand where your pain truly comes from. Daughters of narcissistic mothers can almost always draw their pain points back to their mothers, as their mothers tend to infect every area of their lives with toxic words and beliefs. You might go so far as to find that the pain you experience from being too afraid to go for it in your dream career is because your mother taught you that you were not worthy enough or smart enough to do anything meaningful with your life.

It can be painful and even shocking to realize that much of the pain you have experienced in your life fall back on the way your mother treated you and the way she taught you to believe about yourself. With that being said, you need to get honest about where your pain is coming from and what started this pain in the first place. Every time you feel any level of pain in any area of your life, especially emotionally or mentally, you need to pause and look at where this pain truly comes from. Pinpoint the moment in your history that this pain started, and do your best to remember everything about the situation that caused this pain. Also take note of any subsequent events that reinforced this pain, as these recurring events will have encouraged the pain to become even worse and will also need to be healed.

Knowing where your pain comes from down to the exact moments that caused it and reinforced it is going to help you move forward from them. Now, rather than internalizing the beliefs your mother fed you in those moments, such as you not being good enough or capable enough, you can recall the truth. You feel this pain not because you are truly not good enough, but because you were bullied by your mother into believing that to be true. With this in mind, you can start to heal from the abuse of your mother, rather than staying in the mindset of constantly trying to "fix" yourself to be a better person because your mother has to lead you to believe that you are somehow broken.

LABEL THE PAIN YOU EXPERIENCE

Individuals who have become victims of narcissistic abuse often have a hard time identifying their exact emotions. Throughout your life, you have been consumed with attempting to understand and navigate your mother's emotions, to the point where you may not have any clear understanding as to what emotion feels like within yourself. I know for myself the only emotions I ever recall experiencing until I began healing from my relationship with my mother were anxiety and anger. I would constantly switch back and forth between the anxiety of what abuse would come next and the anger at my mother for not being able to be a better parent to me. At that point, I had no idea that I was also experiencing grief, fear, guilt, sadness, disappointment, overwhelm, trauma, and many other emotions that link to being involved in an abusive mother-daughter relationship.

Learning how to properly label your emotions is going to help you start to know and understand yourself better, and it is also going to teach you to think about yourself. Until now, you have not had much time to think about yourself because you have always been thinking about your mother and what your mother's needs are. Being in that position has likely been incredibly painful, and has led to you feeling your inner sense of self-neglect because you have let yourself and your life be overruled by your mother and her never-ending needs.

I found that the more I learned to properly label my emotions, the more liberated I felt. I began to understand every single emotion I was experiencing, and I was able to truly address those emotions in a way that was more appropriate to what they were. I was also able to start feeling normal for perhaps the first time in my entire life because I realized that my emotions were a natural response to my abuse. I no longer felt like I was bad or wrong for feeling any of these things, but that I was incredibly healthy and normal for having these emotional responses to my upbringing. For the first time in my life, I felt like I had a hope of being someone other than the abused daughter of a narcissistic mother.

The best tip I learned to help me identify my own emotions was to use a wheel of emotions. My therapist gave me this tool to help me identify my emotions, and it works by starting in the center to identify the most basic emotion you are feeling, like "sadness". For example, I was feeling sad because I feel like my mother does not love me, and that made me sad. Then, you follow the wheel out to properly identify what type of sadness you are feeling, until you reach a more definitive emotion, like feeling inferior or isolated, both of which were true for me. When you begin to have these exact emotions that describe how you feel, it becomes clearer as to why you feel the way you do and your ability to start allowing yourself to process these emotions becomes easier, too. Now, rather than feeling confused and overwhelmed by your emotions, you feel aware of and understanding in them.

Keep A Journal of Your Feelings

I strongly advise you to keep a journal as you go through this entire experience and that in that journal you regularly write down how you are feeling. At least a few times per day, and every time you feel an overwhelming influx of emotions, stop to identify what exact emotions you are feeling. You may find that you are being overwhelmed by just one emotion, or you may find you are being overwhelmed by many. There is no right or wrong answer, simply write down everything you are feeling and the symptoms of those emotions that you are feeling.

Then, write down where that emotion comes from, and how you can navigate that emotion now as an adult. Part of navigating that emotion now will likely include accepting that it exists, searching for healthier ways to express that emotion, and allowing yourself to work through it so that you can fully release it.

A real example from my emotion journal looks just like this:

"I feel jealous because my friend Judy has a mother that is so attentive and caring. She calls her every week and they share a great relationship with each other. I know this comes from my mother never sharing that with me, and the realization that we never will. This jealousy makes me feel nauseous, angry, and like crying over what I will never have. I am choosing to accept that this is my reality, and I am going to use this acceptance to help me move through this jealousy. I know my broken relationship with my mother is not my fault, and the feeling will not last forever, and I am ultimately happy to know that Judy gets to share such a great relationship with her mother. It makes me happy to see that this type of love exists in her life, even if it does not exist in mine."

Taking the time to write like this on your own emotions will help you see what your emotions truly are, validate them, and healthily process them. This way, you can move through them and release them, allowing you to make room for more enjoyable emotions like gratitude and acceptance.

Chapter 10

Learning to Meet Your Own Needs

Part of healing from your mother's narcissism is learning to meet your own needs. The average person learns to identify and meet their own needs as a child, but daughters raised by narcissistic mothers do not have the privilege of doing this. Rather than spending time learning how to identify and understand their own emotions, they spend their entire childhoods learning to identify and understand their mother's emotions. Of course, they never succeed because their mother's emotions are so volatile and ever-changing, but they will do their best to try, anyway.

As an adult, this behavior likely pours into your other relationships with you naturally thinking about everyone else before yourself. I did this to the point where it was so automatic that I did not even realize I was doing it until it was done and then I realized my own needs had gone unmet. Then, I would grow incredibly frustrated and resentful to the people in my life for not choosing to see and meet my needs in the way that I had been doing for them. Of course, they were not doing this because this was not a form of healthy or normal behavior, but I had no idea that this was a behavior that was unique to me as a result of my abuse. Not until I started to look into my symptoms and heal from my narcissistic mother, anyway.

Learning how to heal from your mother's abuse and function normally as an independent adult requires you to learn how to see, and meet, your own needs before anyone else's. You have likely heard the phrase before, which states: "you cannot serve from an empty vessel" by Eleanor Brown. When I heard this, it resonated deeply, yet I was not yet aware of how to put this advice into action in my own life. It took a lot of time learning how to discover what my needs were, and teaching myself to identify these needs in a timelier manner and communicate them with assertiveness before overwhelming myself with the needs of others.

As you continue to heal, you need to learn how to make this switch, too. You need to learn how to identify what your needs are and assertively communicate them before you agree to take on anything that would defy your own needs.

Learn How to Identify Your Needs

I found that the biggest secret to identifying my own needs was tapping into my inner dialogue and using it intentionally. In the past, I had ignored my inner dialogue to attempt to consider what my mothers' needs were, which meant my thoughts always revolved around her.

My inner voice sounded a lot like this:

- "What does my mother need?"
- "What will make my mother happy?"
- "How can I keep my mother from getting angry?"
- "How can I please my mother?"
- "What can I do to keep myself from being neglected?"

Every single thought revolved around my mother. When I moved out of her home, the thoughts like that continued, both around the topic of my mother and the topic of anyone else I cared about. Like you, I was taught that respect and love were shown by abnormal levels of appreciation to the point where the person you respect and love is almost the only thing you think about.

Except, that's not healthy.

I began to tap into my inner dialogue and intentionally change my thoughts as I went through my day. Every time I found myself asking about how I could serve my mother, or someone else, I would instead ask how I could serve myself. This would require me to know my own needs, which meant it started the inner dialogue around identifying those needs and understanding them.

At first, identifying your own needs is going to feel strange. You might find yourself feeling guilty and selfish for thinking about yourself, and you might feel bad or wrong for not thinking about others. You may think that they are going to believe you do not love them or care about them if you do not put them first before yourself, because you have been taught that this is the only true way to show your love for someone else. This is not true, and the reality is that most people are not going to want to have you put their needs and desires before yours. In many ways, this makes it feel like you are putting too much pressure on them or attaching them into an unhealthy way, which leads to them not wanting to be around you as much. Alternatively, it could drive you directly into the path of more narcissists and abusers because you are already perfectly groomed to be taken advantage of by them.

As you continue to have this inner dialogue, however, you will start to understand that your own needs are separate from the needs of others. You will come to realize that there is no reason for you to feel responsible for other people's needs because meeting their needs is their responsibility. Meeting your needs is your own (and only) responsibility. And, of course, if you have a younger child of your own then meeting their needs is your responsibility to, to the point that is appropriate for their age.

CREATE RITUALS FOR FULFILLING YOUR NEEDS

Something you can try that might help you begin to take care of your own needs first is creating rituals or habits around how you fulfill your own needs. Rituals that are designed to help you fulfill your own needs can help you turn this into a habit, rather than something that you have to fight to remember to do. It can also make fulfilling your habits feel more fun and enjoyable, making it easier for you to convince yourself that it is worth your while.

The three rituals I keep for myself that help me fulfill my own needs include a ritual for identifying my needs, a ritual for daily stress management, and a ritual to put myself in a positive mood whenever I am feeling down. My ritual for identifying my own needs is truly a simple conversation I have with myself that reminds me to check in and understand myself and my own needs before I commit to anything. My ritual I have for stress management is a bath with a candle, and some time spent feeling my emotions and validating myself. I use this one a lot after my mother has attempted any abusive cycle, as it helps me remember that I am worthy and valuable, despite how she may attempt to make me feel. Lastly, my ritual for helping myself feel better is to put on my favorite music and dance to it for five to ten minutes, no matter how I am feeling when I start. This helps raise my energy and put me in a good mood every single time.

You can decide what rituals you need in your own life to help you meet your own needs and feel better continually. I strongly recommend making them something that feels fun and fulfilling for you, as this will help you genuinely honor your own needs and start taking care of yourself. Even starting by learning to fulfill a few needs at a time can help you grow and have a better experience with managing your own needs first over anyone else's. As you begin to make this a more positive and enjoyable experience, you will find it becomes easier over time.

Permit Yourself to Come First

Although this entire practice of learning to fulfill your own needs has involved you learning to put yourself first, you must pause and give yourself intentional permission to come first. Giving yourself verbal permission to choose you over anyone else is a powerful opportunity for you to take this from a "good idea" and turn it into an actual practice that you are going to use in your life. When you commit to taking things and turning them into an action plan, your life begins to change much faster.

A great way to start permitting yourself to come first, and a great way to remind yourself that you have done so, is to create a mantra around permitting yourself to come first. I made mine as simple as possible, with it saying: "I give myself permission to come first." I repeat this to myself every time I notice I have a need that needs to be met, and then I go through the process of understanding what that need is and how I can reasonably meet it at that very moment.

CHAPTER 11

SETTING STRONG BOUNDARIES

As you might have guessed, part of being involved in a narcissistic mother-daughter relationship is not knowing how to assert boundaries. Overstepping natural boundaries and making you feel like you had nothing that truly belonged to you or that was yours was one of your mother's ways of making you put her first over everything in your life. By doing this, she was able to essentially hijack your brain and make you think about her and her needs over you and your own at all times.

Not having learned boundaries in your childhood can lead to you finding yourself in many situations as an adult where you struggle to assert your boundaries. This very experience may have affected nearly every human interaction you have had in your life until this point. Situations with your other family members, friends, lovers, co-workers, bosses, and even acquaintances may have all lead to you being taken advantage of or treated poorly at one time or another because you did not know how to assert your boundaries. Furthermore, you might have felt at various times that learning to do so would be rude, as you would be directly showing a sign of disrespect or ungratefulness by learning to assert boundaries. You may not realize that there is such thing as positive boundaries that can be asserted politely, or that these boundaries are something that you are allowed to exercise in your life.

It is time for you to begin learning how to assert your boundaries in a polite yet stern manner that will ensure that anyone who hears you asserting your boundaries know you mean business. This will help you stop over-giving and being taken advantage of, and it will prevent others from seeing you as someone that can be taken advantage of. It will also help you finally put an end to the painful question of: "why is everyone else being treated better by this person, except me?" A question that is commonly asked by victims of narcissistic abuse who have yet to learn how to assert themselves and their boundaries.

Learn How to Say "No"

One of the first things you need to teach yourself when you are learning to set and assert boundaries is the word "no." As a daughter of a narcissist, you have been taught that "no" is not only an unacceptable word but also a dangerous word. In your childhood, you were likely punished in major ways, including being berated or neglected, for using the word "no" with your mother. She may have even smeared you to other people by painting herself as the victim any time you attempted to say no to her. Whatever measures she took, you have likely learned that saying no is dangerous and not ideal in any circumstance, ever.

What you did not realize as a child was that almost no one reacts this way. Having an excessive or exaggerated response to the word "no" is not normal, or healthy, and it is also not common. Most people will not be abusive to you if you say no to them. I'm guessing that you may have realized this at various points in your life, but that you still find yourself feeling overly sensitive every time someone shows any signs of disappointment when you say no. This is because you have been taught that their feelings are your responsibility, through the unhealthy expectations of your mother.

In learning to say no, you need to learn that it is A) appropriate and healthy to say no, and B) it is not your responsibility to manage other people's emotions. You need to exercise your new skills of fulfilling your own needs and use the word "no" as a way to do so consistently, whenever you need to. By practicing using the word "no" anytime you find yourself in a situation where you are being asked to do something that makes you feel as though your own needs are going unfulfilled, you will start reinforcing both your needs and your boundaries. This will make doing both much easier going forward.

BE DIRECT, YET POLITE WHEN ASSERTING YOURSELF

A big fear that daughters of narcissistic mothers have when it comes to setting boundaries is the belief that no matter how you do it, asserting your boundaries is a sign of ignorance and a lack of love and respect. When you learn to assert yourself, it can be helpful to realize that asserting your boundaries is a sign of love for both yourself and the other person, as you are valuing the healthy relationship you share with them. The only way you can do that is if you value yourself and your health, which comes with asserting your needs and your boundaries.

You can help yourself begin to see boundaries as a positive thing by realizing that there are ways to politely assert your boundaries that are still direct and effective. By using the direct yet polite approach, you can start to rebuild your beliefs around what boundaries are and how they serve you in your life.

A great direct yet polite way of asserting your boundaries would be as simple as saying: "No thank you" when someone asks if you want to do something. If they continue to pressure you, you can say "I have asserted my boundary by saying no thank you, please stop." If they continue further, you can say "I do not feel that you are respecting me, I am going to end this conversation now." This is not disrespectful but is instead a sign of you loving yourself and asserting your boundaries and needs. In this, you did not berate or disrespect anyone, but you did show love and respect to yourself, which means that it is a positive and healthy example of setting boundaries.

It is important that when you do issue a consequence of what will happen if your boundaries continue to be ignored that you follow through. Not following through on your consequences will result in people not believing in you and your boundaries, which will lead to them continually overstepping your boundaries no matter how hard you attempt to assert them. This is because you have taught them that your boundaries ultimately do not mean anything and that you can be taken advantage of, despite you trying to avoid this. You need to be assertive and remain assertive, including by following through on your promised actions, to show people that you are not willing to be taken advantage of. In doing so, people will either leave you alone or learn to treat you in a way that is more appropriate to how you desire to and deserve to, be treated.

CREATE A SYSTEM OF PERSONAL RULES (AND FOLLOW THEM)

Creating boundaries in your life when you are not used to having boundaries in your life can be a difficult experience. At this point, you may have absolutely zero boundaries that you are regularly asserting in your life, which may make you feel like you have to start from scratch because, in a sense, you do. It can be overwhelming to look at all areas of your life and realize that your lack of boundaries leaves you vulnerable and susceptible to people's mistreatment. You could start to feel like there are a lot of areas of your life that need to be addressed to avoid letting people, like your mother, take advantage of you any further.

When I was first learning to assert my boundaries, I chose to create a system of personal rules and then apply them in every situation where boundaries were needed. These personal rules helped me identify my needs and desires, and then assert my boundaries to help me ensure that my needs and desires were fulfilled. They also helped me keep my boundaries with myself by not breaking my rules and letting people, including me, take advantage of myself. This way, I was confident that I would treat myself with the level of respect that I deserved and that I would ensure everyone else would too if they desired to be in my life and presence.

Creating personal rules for yourself can help give you a guideline of healthier practices to follow in your day to day life. This way, you can step away from the toxic behaviors your mother taught you when you were growing up, such as not to have or assert your boundaries, and start living a healthier life.

The rules you create should reflect who you are, what you desire from life, how you desire to get it, and what makes you feel your best in your everyday experience. They can involve how you are willing to treat yourself and how you are willing to let others treat yourself, how you are going to make decisions that serve you, and anything else that genuinely helps you live a better quality of life.

Some of my own life rules include:

1. I will not let anyone take advantage of me or overstep my boundaries.

2. I will always put my own needs before anyone else's.

3. I will not do anything that does not ultimately make me feel happy, loved, or supported.

4. I will only allow healthy, non-toxic relationships to exist in my life.

5. I will not let anyone minimize or devalue the experiences I have.

6. My perception of reality is my truth, and no one can tell me otherwise.

7. I will validate myself and not seek validation from others.

8. I will not place my value on the perception that others have of me.

9. My own happiness and wellbeing is the most important thing in my life.

10. I am not required to go out of my way to fulfill others' needs.

You should make your list of rules in your journal that you can use and reflect on to help you begin identifying your needs and asserting your boundaries. I suggest reading that list every single day to remind you of who you are and what you will and will not allow happening in your life. Then, to the best of your ability, uphold all of those rules on a day to day basis. The more you do, the more empowered you will feel to take control over your life and stop letting other people take advantage of you.

Chapter 12

Choosing Forgiveness

I'm not going to lie: I have declared that many parts of this process are going to be challenging because they are. Healing from narcissism is not easy in any sense of the experience, and virtually every step is going to have a level of difficulty to it as you face the reality of what you have come through and teach yourself to be a healthy, functioning member of society.

The forgiveness part of the healing process offers its unique type of pain and confusion, however, and I want to highlight that before we dig into it. When you reach the point where it is time for you to begin forgiving people for the pain, they have caused you, and yourself for the ways you have contributed, you reach a point where you truly need to commit to changing your perspective. At this point, you need to accept that you are removed from the situation and admit that you are ready to see things differently. You are ready to see things with a loving, compassionate perspective that encompasses having love and compassion for yourself, too.

Forgiveness is often seen by victims as a way to justify what the abuser has done and, in some way, declare that what they did no longer matters, or that the pain no longer exists. Staying in this mindset can keep you feeling troubled and confused around the topic of forgiveness because it may feel like you are attempting to erase or justify everything your mother has done to you in your life. Another angle here is that if you decide everything, she did in the past is now "okay" then in some ways you are admitting that her repeated abusive behaviors are somehow "okay", too. After all, how can something be considered okay in the past but not okay in the future? This perspective is one I know all too well, and it is one that can add to the confusion and inner turmoil around trying to forgive your mother.

As you ponder on the topic of forgiveness, you may also start to feel an intense sense of overwhelming. Maybe you feel that some things will be easier to forgive, but others will not be. Understand that forgiving your mother does not have to be done all at once, as blanket forgiveness does not have to be issued. You can choose to forgive one thing at a time as they come up, while also choosing to forgive the experiences you have had in your life until now.

One last toxic perspective I want to address is if you feel like you have something new to forgive every day and so you are in a chronic state of having to attempt to forgive your mother for everything she continues to do. If this is how you are feeling about the topic of forgiveness, understand that in this experience you are not truly forgiving your mother but instead you are using forgiveness as a way to enable her to continue behaving in a toxic manner. You are not required to forgive her and enable her to continue treating you in the same way that warranted forgiveness in the first place. You can exercise forgiveness with boundaries so that you can have your needs met and distance yourself from her abusive behaviors. That is a choice you can make, and it is also a choice I recommend that you do make. Doing so will prove to you that forgiveness is truly an act of self-love, and not an enabling or minimizing behavior that is meant to take away your right to being outraged and hurt by how you have been treated.

With all that being said, forgiveness is a crucial step in healing. Forgiveness is a step that enables you to choose to no longer give power to the painful things people have done to you, including yourself, and instead give the power back to yourself so that you can heal those wounds. When you forgive someone, you decide that you are going to accept the situation as it is and place your energy on becoming aware of your pain so you can heal it, rather than placing your energy and hate and revenge toward the person that hurt you. Forgiving is a way of letting yourself off the hook so that you can begin to make changes in your life by asserting loving boundaries and moving on from the situations that hurt you in the first place.

Learn to Forgive Your Mother in Your Way

The very first phase of forgiveness you need to begin with is learning how to forgive your mother. Your mother is the person who hurt you first, and who likely hurt you deepest. She is the person who filled your heart with disappointment, rejection, abandonment, fear, neglect, sorrow, grief, guilt, shame, and many other painful emotions that you live with every day. She is the one who ultimately let you down and failed to provide you with the safe, loving, and nourishing childhood that every child deserves to experience, and requires to thrive and grow into a healthy, functioning adult. The reason why you are here right now reading this book and healing from this in the first place is because of your mother and her behaviors.

Your mother is the source of much of your pain, and she has used this pain to feel powerful over you for most of your life. By keeping you in this state of pain, and by you allowing it, you are letting your mother continue to control you even when she is not around to control you through her active words and actions. Her voice rings through your head and reminds you of all of her moments of cruelty and keeps you down, preventing you from being able to fully heal.

Choosing to forgive your mother in your way means finding the balance between forgiving her for not being able to be a better mother to you, and setting strong loving boundaries that prevent her toxic behaviors from impacting you anymore. How this looks exactly is going to depend on what feels right for you, and where that balance lies for you between forgiveness and boundaries.

The reason why forgiveness specifically is important here is that when you switch into a perspective of having forgiveness for your mother, you also have compassion for her. This compassion supports you with releasing the feelings that are associated with your pain so that you can let go of that pain as you begin to heal through it. As a result, you are not creating more pain for yourself every time you think about your mother or the experiences you two have shared, or not shared. Having healthy boundaries in place will ensure that this compassion is not used against you. For example, with boundaries in place you are less likely to be compassionate to the point where you justify her behaviors and allow her to continue treating you in this poor manner. This way, you can release the pain without exposing yourself to more abuse and trauma in the process.

Forgive Yourself for Your Experiences

Another person you need to forgive in this experience is yourself. You need to understand that for most of your life you were young and unable to speak up for yourself, and during those years your mother wrongfully took advantage of you and taught you not to speak up for yourself. This way, as an adult, you would be less likely to speak up and end the cycle of abuse which would ultimately enable her to continue behaving this way.

You are not responsible for how your mother treated you, or for how you were groomed as a child to comply with her unreasonable demands as an adult. It is not your fault that your mother has a mental condition, or that this mental condition affected you and caused you to adopt unhealthy behaviors and coping methods at various points throughout your life. Even though you may feel ashamed of how these behaviors or coping methods have affected your relationship or reputation with others, it is not your fault that this happened. When it was happening, you had no idea that this was not the "appropriate" way to act, because to you this was the only appropriate way to navigate the incredible amounts of pain and abuse your mother was exposing you to daily. Most people who judged you had no idea what was truly going on and those who did likely had no way of comprehending how bad it was due to the way your mother expertly hid her abuse through grooming you and others not to see it when it happened.

Accepting this all to be true is a powerful opportunity for you to forgive yourself, which will help you stop blaming yourself and punishing yourself for the ways you have acted in the past. You can accept that you did not know better at the time and that the way people treated or viewed you due to your behavior was not because you were bad or wrong, but because neither of you truly understood what was happening. You had no way of knowing that your behaviors would be seen as abnormal or strange by others, and they had no way of knowing what you were truly going through that caused you to behave that way. It was not your fault that your mother treated you that way or that others did, and so you need to forgive yourself.

As you forgive yourself, learn to have compassion for yourself and all that you have been through, and all that you did to cope with it. Also learn how to have boundaries with yourself so that you no longer allow yourself to continue punishing or blaming yourself for these things, or otherwise reinforcing the pain and trauma through your own words, thoughts, or behaviors. Learn to release and let go, and set a boundary that permits you to become a new person, with new coping methods and behaviors, today. Commit to moving forward more healthily, and watch how much this level of forgiveness transforms your relationship with yourself, and your life in general.

CHOOSE TO FORGIVE THOSE WHO DIDN'T BELIEVE YOU

The final area of forgiveness you need to focus on is forgiving those who did not believe you or what you were going through. People who chalked your pain or abuse up to being a "typical mother-daughter relationship" or to you being overdramatic had no way of knowing what was truly going on because your mother expertly hid it from them. Or, maybe they experienced similar traumas from their mother and have not yet come to terms with it or learned to view abuse for what it truly is. Maybe they carry their unhealthy perspectives on what a healthy relationship should look like, and how love and respect are shown to others.

When it comes to forgiving and healing from your narcissistic abuse, forgiving others will also help prevent you from holding them to blame for what you went through. It will also help you stop punishing them in your mind for not being able to help you or validate you, which will support you in ending the cycle of painful emotions relating to your mother's abuse. Make sure that in forgiving others, you learn to have compassion for them while also having boundaries with them. Do not forgive others and allow them to continue minimizing the pain you carry or the level of abuse you experienced, or otherwise downplaying what you went through. After you have forgiven them, assert boundaries around what you are and are not willing to discuss with them, and reserve your right to withhold discussions about your mother with anyone who cannot respect what you went through. You are not obligated to let people judge or pass opinions to you based on what you went through, no matter who that person may be.

Chapter 13

Working With Therapists

Whenever you are healing from any form of abuse, I always advise that you work with a qualified therapist who can help you completely heal from the abuse you have faced in your life. Especially in a situation that is as complex as healing from narcissism, having a therapist can help you work through the challenging emotions and realizations and have compassion for yourself. They can also support you in developing healthier coping methods and self-care routines while keeping you accountable in your commitment to living a healthier life.

I always recommend the route I went when I finally removed myself from my narcissistic mother's abusive cycle, which was to combine self-help with professional therapists. I read as much as I could, surrounded myself with supportive people, and did everything I could to inform and educate myself on what I was going through and how I could successfully get through it. Then, I also hired a professional therapist who could help me in ways that I could not help myself and that would not available through my loved ones or books. My therapist has helped me understand my unique situation, create custom strategies for healing and coping based on my personal needs, and has ultimately supported me in feeling safe and comfortable during the entire experience.

Hiring a therapist can seem scary, especially if you have had a negative experience with one. Personally, in my childhood, my school connected me with a childhood therapist who ended up speaking with my mother behind my back which resulted in her believing I was lying and in need of help for being a chronic liar, rather than a child that was being abused. It took me some time to heal from this and confide in a new therapist, but I am grateful that I did. My therapist now in adulthood is private, only communicates with me, and is fully committed to helping me get through my challenges relating to my mother and many other things in my life at this point. Remember that you are now an adult and that your mother does not have the power to negatively influence your access to help, and that you are entitled to receiving the help that you need.

SELF-VALIDATE YOUR RIGHT TO SEEK HELP

Right now, is an excellent time for you to practice self-validation as you validate to yourself that you have a right to seek help and that you are deserving of the help you desire. Use this as an opportunity to prove your commitment to yourself and your needs, to assert boundaries in your mind to the thoughts that tell you that you do not need help, and to forgive yourself for your fears around help. You can even use this as a time to label and work through the emotions you are having around the idea of hiring a therapist in the first place to seek help.

Whatever it takes for you to commit to finding and working with a therapist and getting help, I strongly advise you do. While I cannot force you to go see anyone, I also cannot stress enough how much having a therapist has transformed my ability to completely move beyond the trauma my mother inflicted upon me throughout my lifetime. I meet with my therapist monthly, and I always meet with her before and after I see my mother as this ensures I stay on track with my healing. My therapist has helped me hold myself accountable in countless experiences, and when you are healing from abuse, especially abuse that started before you could even talk, this is so important. Please consider getting yourself a therapist to help you through this.

FIND A TRAUMA-INFORMED THERAPIST

In finding a therapist, I want to point out that you need to find a trauma-informed therapist. These days, many therapists make an effort to be trauma-informed which means that there should be no shortage of therapists available to you to help you with what you are going through. With that being said, do make sure that when you are looking for a therapist you ask them what sorts of the trauma, they have helped people heal from, and what their philosophies on healing from trauma are. You can also ask them about their experience with narcissism specifically. Knowing that your therapist understands your unique type of trauma and what you might be going through can help assure yourself that they are going to believe you and be helpful to your healing experience. This may help you feel more confident in actually attending your therapy appointments and opening up to your therapist, too.

CREATE A SENSE OF SAFETY IN YOUR CLIENT-THERAPIST RELATIONSHIP

Always make sure that when you work with a therapist you pick one who helps you feel safe and supported right from day one. It might be challenging to tell if you are particularly afraid of visiting a therapist in general, but typically you will know because you will speak with a therapist who seems to help you feel better. You want to pick a therapist who helps you feel more comfortable and supported from the start, as this is a therapist that you are likely to develop a good relationship with. If you find you have a therapist who you do not feel comfortable with, recognize that this is likely a mismatch between you and your therapist's personality and not evidence that therapy will not help you.

If you do have fear around visiting your therapist, even if you think that fear sounds silly or strange, do not be afraid to open up about this. You could even make this your first area of focus so that you can test the waters to see how your therapist responds to your emotions and your needs. In many instances, sharing this will help you feel more confident and will allow your therapist to understand your needs while also showing you that they are there to help you, not judge or hurt you.

Lastly, I recommend keeping the topic of your therapist away from your mother unless you truly feel the need to tell her. I did not tell my mother about my therapist for years, and I have never discussed why I am in therapy with her. Telling your mother that you go to therapy, or telling her that you go because of her, could expose you to being abused by your mother for your choice which could compromise your willingness to continue going. Some things are better kept to yourself, and this is often one of those things. You can do so by asserting the boundary that you are not required to tell your mother everything can be incredibly helpful in establishing a sense of security in your client-therapist relationship.

Chapter 14

Evaluating Other Relationships

An unfortunate experience that many victims of childhood abuse from a narcissist have is that they begin to experience abuse from other people in their lives. These other people seem to sense that they can withstand abuse and that they will do so without asserting boundaries or walking away because they truly do not know how to protect themselves. As a result, they become the victims of all sorts of abuse throughout their lives, to the point where they may wonder why they are the "chosen ones" to be victims in so many different ways.

If you have been involved in an abusive relationship with a narcissist in your life, there is a good chance that you have experienced abuse, or at least abusive behaviors, in other areas of your life, too. Being groomed from childhood to remain open and vulnerable to your mother's abuse means that you are more vulnerable to other abusive people's abuse, too. Other abusers will recognize that you are incapable of protecting yourself and will use this to their advantage by turning you into their victim as well, leaving you exposed to multiple forms of abuse.

Being exposed in this way and enduring the abuse from many people can leave you feeling many different things. You will likely begin to experience an incredibly low sense of self-worth, self-esteem, and self-confidence as it feels like many people in your life are harming you. You may also struggle to believe that there are any good people in the world because you are constantly wrapped up in cycles of abuse with various people who leave you feeling like the only thing that exists is pain and abuse.

Furthermore, you might carry that lack of trust into any positive or healthy relationship that does begin to form in your life, resulting in you self-sabotaging that relationship and losing it. When this begins to happen, you may even start blaming yourself for why you are not in better relationships, which can further worsen your experiences in the relationships in your life. You might begin to think that your inability to maintain healthy relationships means that it is entirely your fault that you are not in any healthy relationship, and that you deserve to be trapped in these abusive relationships you are a part of. Of course, this is not true, but this is a highly common cycle for abuse victims to go through, which can lead them to stay in unhealthy relationships even if they think they know better.

REFLECT ON YOUR EXISTING RELATIONSHIPS

The first thing you need to do when evaluating your relationships is reflected in your existing relationships. Consider who your top 2-5 friends are, or who you spend most of your time with, and reflect on the relationships you share with these people. Pay attention to important things such as how you feel around these people and how they treat you. Consider them against your list of personal rules, particularly the rules around how you are willing to let other people treat you, and see how they compare. Are they generally treating you in a way that is healthy and kind? Or are they treating you in a way that violates your new personal rules, and boundaries?

In addition to considering the other person, consider how you behave in these relationships. Are you behaving like a healthy individual as a part of these relationships, or are you contributing to the relationship in a negative, unhealthy manner? You might realize that you are behaving in a way that results in you taking responsibility for their emotions and behaviors, which is unhealthy. In this case, their unhealthy behaviors may not necessarily be a reflection of them, but instead, it may be a reflection of how you have taught them to treat you. In doing so, you may have allowed them to treat you poorly and they may have gone along with it without truly realizing that this is what they were doing.

Be objective when you are reflecting on your relationships so that you can place responsibility where it belongs. In other words, you need to take responsibility for your actions, but do not take responsibility for theirs. Even if their actions are a response to your actions, hold them accountable for how they treat you so that you can contribute to a healthier dynamic in your relationship with them.

LEARN TO CREATE HEALTHIER RELATIONSHIP DYNAMICS

Upon reflecting on your relationships, you are also going to want to learn new ways to engage in relationships in a healthier way. You want to make sure that in the relationships where healthier dynamics can be achieved, you work toward achieving them. You can do so through asserting your boundaries, communicating your needs better, and taking the necessary steps to put yourself first rather than always putting others first. It is also a good idea to let the person you share a relationship with understanding what you are going through, how it has affected you and your relationship with them, and what you are going to do to help improve the relationship itself.

As you work toward improving your relationships and engaging in healthier relationship dynamics with people, one of three things is going to happen in each relationship. The person is either going to agree and be supportive, agree after some resistance or disagree or remain resistant to the changes altogether.

When the person agrees and is supportive, you can feel confident that this is going to be a relationship that will be able to foster a healthier dynamic. Chances are, this is someone you can maintain a relationship with for a long time and they will go on to be wonderfully supportive and helpful of all of the changes you are going through. They may also be great people to confide in during your healing journey, depending on who they are and how your relationship with them looks.

When the person agrees with resistance, this could mean a few things. They may be someone who has a hard time changing in general and who genuinely require a few reminders to fully change their behaviors and stop treating you poorly. Or, they may be reluctant to change because the relationship's current dynamics are in some way serving them. Lastly, they might resist the change altogether despite seeming to agree with the need for change because they are benefitting from taking advantage of you and your coping methods. In both of the latter scenarios, you want to seriously consider the future of your relationship with this person as they are unlikely to be healthy for you to keep around. It may be ideal to minimize your contact with them or end the relationship altogether if they never truly change. Remember, no matter how they have treated you in the past, or how you have let them treat you, it is not your obligation to let them continue to treat you poorly now. You have a right to change your mind, require better treatment, and expect change. If they do not change, it is a sign that they do not respect you, even if they claim they do or that they are trying to change. Simply put, someone who wants and cares about change will change, especially if they learn that their existing behaviors are hurting you.

If you come across someone who is completely resistant and refuses your requests or refuses to admit that change is necessary, there is a chance that this is someone abusive. They are likely gaining some form of benefit from you allowing yourself to remain the victim and letting them take advantage of you, and they are completely unwilling to change their behaviors. This is not a person to keep in your circle, no matter who they are because they can be toxic and damaging to you and your self-esteem. Keeping them in your life can impact your ability to move forward and heal, which will directly oppose what you are trying to accomplish. As hard as it might be, you need to let them go and move on in your life.

BUILD A HEALTHIER SOCIAL CIRCLE

Assessing and restructuring your existing circle can be a challenge. For me, doing this resulted in me losing many of my friends and even close relationships with my family because they were simply unwilling to treat me with greater respect, and I was unwilling to accept disrespectful treatment any longer. Part of me staying firm in my boundaries meant ending these relationships and moving on with my life.

It might feel like if you terminate your relationships you are going to be completely alone, especially if this is something your mother has lead you to believe would be true. After all, forcing you to believe that being trapped was the only way to be loved is how she kept you in her cycle of abuse for so long. Understand, however, that there is an entirely different perspective that you can take around this circumstance that might help the process be easier for you.

Understand that terminating relationship and upholding your boundaries does not mean that you are going to be alone. It is not a sign that you are lacking in love, support, compassion, or attention from people in your life. If you let go of these unhealthy relationships, you are not going to be completely forgotten about and abandoned by the world, even if it truly feels that way in your heart. You are not going to be any worse off than you are right now. You are going to be better off.

This is not just because you will no longer be exposed to their abuse, even though that is a big part of it. However, an even bigger part is that when you are not wasting all of your energy running in circles and trying to pick yourself up from people's abuse, you have more energy to build healthier relationships in your life. You can focus on meeting new people, people that are completely removed from the abusive part of your life. You can spend time with people who share your interests, and who respect you, and who genuinely enjoy your company and are willing to build a healthy relationship with you. In this, you can even start to explore who you truly are without the influence of abuse when you connect with people who have never known you as a victim.

Building your social circle to include new healthy relationships can help you feel as though you are firmly leaving your past in the past and moving forward. This is a big step in taking control over how you feel, and who you are and reshaping your life without the abuse of others.

As you do build your social circle, make sure you do it at your own pace. This is another great part of your life to include your therapist in, as he or she can help you build these relationships in a way that is not founded on your lack of self-esteem or self-confidence. Instead, they can help you learn to feel empowered and confident in your relationships so that you can genuinely enjoy positive, healthy relationships with new people. You will find that as you do this, and as your relationships get stronger and stronger, your life will continue to improve significantly and you will have more opportunities to feel "normal".

Chapter 15

Protecting Yourself From Abuse

The final piece in allowing yourself space and opportunity to heal from the abuse is knowing how to protect yourself from future abuse. During my healing process, I found that I had an incredible fear around developing relationships with new people, only to find that they were abusers, too. Because many abusers can go undiscovered for so long before finally showing their true colors, I was terrified of trusting in and caring about anyone for fear that it would turn out terribly. I was truly so untrusting of everyone around me.

You might find that you feel the same way. It might be hard for you to fully trust people or let people in because you are terrified of being abused again by someone new. The idea of getting to know someone and care about someone only to have them hurt you in this painful way seems too much to bear, yet it also seems inevitable at times. At this point, you are using your past to shape your vision of your future, and it is looking as though you do not know how to truly protect yourself from people who want to hurt you or take advantage of you. That makes complete and perfect sense, and it is also an incredibly normal way to feel when you have been abused.

Learning how to protect yourself means giving yourself, and your inner child, the key that you never had before. This is the armor you needed all along to protect yourself from abuse and prevent yourself from being harmed by other people. A large number of your fears will begin to drop when you realize that you now have access to the knowledge around what this protection is and how it works, and the willingness to use it. When you realize that you can protect yourself, you no longer have to worry about someone else harming you because you will be able to protect yourself and spot unhealthy relationships from a mile away. Your confidence in yourself will grow exponentially, and as it does your trust issues will start to become less powerful and scary to live with.

You must learn to protect yourself not only so that you can build your confidence, but also so that you can avoid future instances of abuse. Abuse, in any form, is scary and traumatizing. As someone who has already been abused you are especially vulnerable to being abused again, and knowing how to protect yourself is your key to minimizing your vulnerability and exposure in this area of your life.

Learn to Identify Red Flags and Respect Them

The first thing you need to do when it comes to protecting yourself is to learn about what red flags are, and how to identify them in relationships. Red flags are any sign that indicates a relationship is not ideal for you to be in, and they are always true indicators of what you can expect in the future. When you see a red flag arise in any relationship you are a part of, it is a sign to stop pursuing that relationship. You should always, always respect red flags because they never lie.

Some common signs of red flags in relationships include an unwillingness to respect you, overstepping your boundaries even when you have asserted yourself, lying, talking badly about you to your face or others, or ignoring you. Another red flag you should look for, especially with narcissism, is excessive affection early on which can indicate that they are attempting to "love bomb" you –

something commonly done by narcissists. When they love bomb, they are trying to show you that they are a perfect match for you, and then they will quietly work on tearing you down through a constant tug-of-war between showing excessive amounts of love and then abusing you. Not all signs of affection or admiration are love bombing, but excessive ones that seem too good to be true might be.

If you notice someone seems to be using you or trying to change you to fulfill their needs, this is also a red flag. People should respect you as you are and not be trying to change you if they do respect you. As well, if they make you feel inadequate, try to come in between you and the things or people you love, or otherwise, try to isolate you, this is a red flag. There are many additional red flags you can look for, depending on what type of relationship you are considering. A great way to identify what you should be looking for, and what it will look like in real life, is through Googling red flags relating to the relationship in question. This way, you can get a deeper understanding of what you are looking for and how to identify it.

REINFORCE YOUR INDEPENDENCE

Another thing you need to do to start protecting yourself is to reinforce your independence. Continually work on building up your self-confidence, self-esteem, and self-worth through fostering healthier relationships with yourself and with others. The more you can increase your sense of self and your independence, the more you will be able to rely on yourself and trust in yourself. This way, you are less likely to leave yourself exposed to the abuse of anyone in the future.

A great way to continue to reinforce your independence is to get to know yourself better and to have private things, just for you. For example, if you love working out you can make it your private thing to work out regularly. Perhaps for this, you go by yourself and you refuse to invite anyone or cancel your workouts for anyone. Continue doing the things that you do for yourself even when you are in relationships with people, and avoid anyone unwilling to respect your time. Reinforcing your independence both inside and outside of relationships will help you stay stronger in your sense of self, while also minimizing your vulnerability to abuse.

When you do find yourself being exposed to some form of abuse, rely on your independence to help you walk away. Do not be afraid to face the feelings of being alone or abandoned, as you ultimately know this is not true and this will not happen. Reinforce your trust in yourself that you will be safe, loved, and supported no matter what decision you make, and always make the decision that supports your mental, emotional, and physical health.

HAVE A PLAN FOR WHAT YOU WILL DO?

It is always important to have a plan in place for what you are going to do if you find yourself in an abusive situation, or in a situation where you may be in the process of building a relationship with someone that you realize is abusive. This can happen, especially early on in your recovery, and it can be scary to realize that you are in this situation. Sometimes it can take a while for you to catch on, so the situation may have time to escalate before you understand what is happening.

Even if you feel confident that this will not happen, you should have a plan for what you are going to do to escape that situation. Maintain your independence and use your independence to help you remove yourself from the relationship. For example, have a separate set of friends or loved ones you can rely on for support, stay financially independent, and decide how you are going to walk away. Make a plan for how you will assert your boundaries, how you will avoid further contact with this person, and how you will deal with the painful feelings that come up around this.

Having a plan in place helps you feel safer in knowing that if you ever did find yourself in need, you have a clear way to protect yourself and you are not going to be left alone, confused, and trapped again. This is not your childhood and this person is not your guardian: you do have the right to say no and terminate any relationship for any reason, including one relating to abuse or toxic behaviors. That includes your relationship with your mother if you ever find yourself in the abuse cycle again and in need of an escape plan.

Conclusion

I want to take a moment to deeply congratulate you on your willingness to see the truth of your experiences and work toward healing from your narcissistic mother's abuse. Being the daughter of a narcissistic mother is not easy, and recovering from her abuse is not, either. Your decision to put yourself first and take your recovery, and life, into your own hands in this way is a major step in a positive direction and I want you to know that this step is worthy of your celebration.

You are a strong woman, despite how your mother treated you in your childhood. You are not a victim, you are a warrior who is choosing a new path for herself and who is bravely taking action on walking that path, no matter how hard it might be. That is incredible.

Right now, you might be feeling many things. You might be feeling excited to know that there is a future without abuse available to you, or you might be feeling terrified that this is all going to fall apart. Maybe you think it's too good to be true. In either case, I want you to honor how you feel because that is an important part of moving forward.

When I first took big, bold action toward recovering, I went through many different emotions that lead to me feeling like I was going crazy. I would constantly be overcome with huge feelings of relief and excitement, and massive periods of grief and misery. This is all a natural part of the grieving and healing process.

Now, years later, I can honestly say that I have come a long way from the abusive cycles I used to be trapped in with my mother.

She no longer has the power to hurt me, and the words that used to ring through my head like knives cutting me down no longer exist. When I do find myself thinking critically about myself in ways that my mother would have when I was younger, I know that this is a form of intrusive thought and I release the thought and forgive myself for letting it come through. Then, I validate myself and move forward.

One day, maybe soon, you too are going to find that your mother no longer has that power over you. You will no longer feel a constant need to try and please her, and the dreaded fear that comes with realizing that you are trapped in this cycle of abuse. You will feel liberated, confident, and normal. I cannot say when for sure, but I can say with confidence that if you keep working toward your recovery, you can heal from your mother's abuse.

Please stay consistent in your path and continue working toward your recovery every day, as it will take time for you to fully recover. Be patient with yourself, have compassion for yourself, and show yourself as much love as you can. Trust that you are doing your best in every moment and that your life will get better. You have the power to make that happen, and you are tapping into that power, now.

As you continue to move forward with your healing, be sure to continue educating yourself around what to expect, and surround yourself with people who can help you. The more informed you are, and the more supported you feel, the easier it will be for you to truly move beyond your mother's abuse and into your recovery. Keep building up your support system and leaning into them when you need to, and trust that they are there to help you. They care about you, love you, and want to see you succeed. There are genuine, good people in the world, and you can and will find them and surround yourself with them.

I believe in you.

Before you go, I want to ask you one important favor. If you feel that reading Narcissistic Mother. It's Not Your Fault. has helped you in recovering from your mother's narcissism, please take the time to honestly review it on Amazon Kindle. I want to get this book in front of as many people as I possibly can so that I can help more women just like you recover from their abusive situations.

Thank you, and best of luck. You are doing great, and you are never alone.

Self-Love Workbook for Women

The Complete Guide to Finding Your Inner Love, Boosting Your Self-Esteem, and Practicing Self-Care

DESIRÉE SHANNON

Description

To all the women out there, how do you see yourself fitting in this puzzle called life?

At work, you go the extra mile, you give great presentations and you put in that extra effort to make sure each project is perfect. You come in early and help a colleague or you stay back late to make sure those finance sheets are balanced. You are smart, passionate, and dedicated.

At home, you rush back from work, so you have enough time to cook a hot meal for the family, spend time reading with your kids, have enough time for your husband. You spend that few hours after everyone has gone to bed to clean, arrange and prepare meals for breakfast tomorrow. You give so much that there is little time to even do that 15-minute Zumba workout you've been wanting to complete. You barely even have time to call that college friend that is in town for the weekend.

Where do you fit in? Where do you place yourself on the ladder of priorities?

Along the way you compromise your dreams, disregard your health and place yourself at the bottom of the priority ladder. Is there all there is to live for you?

You weren't like this in your 20s. What happened to you?

You can say that life has happened to you but in fact, you let yourself go. You decided that you were not a priority. And that needs to stop. Loving yourself, taking care of yourself, and focusing on your needs extremely important. Only when we take care of our needs and love yourself would we be able to water and love the people around us.

Think of why flight attendants tell you to put the oxygen mask on yourself first before assisting others? It is the same with self-love.

The Self-Love Workbook for Women will help you refocus your needs and priorities. In this book, you will learn what self-love is and you will also learn to make changes in yourself to bring back the focus to you. It is not about being self-centered.

Self-love is about caring for yourself and your needs first, about compassion and emotional intelligence. This book will help you reignite a journey of self-love, especially when you have no clue where to start or how to start. It will give you ideas on what self-love rituals you can do, how you can include them in your hectic lifestyles and above all, how to be disciplined enough to continue reminding yourself that self-love is forever changing and forever adapting journey. There is no end destination to self-love because as our needs and interests in life change, so will our self-love practices.

INTRODUCTION

Welcome to the first step in loving yourself, taking care of yourself and understanding that you alone are responsible for your happiness. By opening this book and reading its chapters, you are already beginning to create a sense of empathy and compassion towards yourself, understanding your strengths and weaknesses and above all creating a life that is both healthy and fulfilling.

This book is all about Self-Love, what it is, what it isn't, what are the fundamentals of self-love as well exercises and things you can do to create an environment that projects self-love not only to yourself but to the people around you.

Are you ready?

WHAT ARE SELF-CARE AND SELF-LOVE?

Let's begin with self-care. Self-care is all about giving or allocating a little time to ourselves in the midst of all the buzz and hecticness that surrounds us. From taking two or three jobs to make ends meet, commuting to and from work, working long hours, waking up early, sending the kids to school, cooking, meeting our professional goals and spending time with loved ones, we also need to give ourselves some time to feel renewed, to focus on our needs as an individual as well not feeling overwhelmed and stressed by the end of it all.

So, to define self-care, it is the act of taking care of yourself mentally and physically. It is all about making sure you have time to feel at peace, to feel centered and to feel serene. Self-care can be anything from spending a few minutes at night to read, or to meditate, to watching one episode of your favorite sitcom to even indulging in that chocolate mousse cup you've been saving for your cheat day.

Self-care is about that self-awareness that you must, you need, and you should allocate time (even for a bit) to yourself so that you can grow and develop. It can even be 10 minutes of jade rolling your face or 10 minutes with a face mask on. For some people, baking a cake is self-care.

Self-love on the other end of the spectrum focuses on cultivating acceptance and gratitude towards yourself both emotionally and physically. It focuses more on our minds and the way we think about ourselves and it can be anything from letting go of our negative mindsets, practicing more self-compassion, being mindful as well as working on physical aspects of ourselves to live a healthier lifestyle.

Self-love is about loving ourselves unapologetically.

Self-care is about taking time to feel good in our skin.

While both self-care and self-love are different, they are both needed to lead a fulfilling and healthy life.

With self-care, you can look forward to:

- maintaining a healthy relationship with yourself where you do not berate or beat yourself up just because you did not accomplish something.

- look forward to having time to ourselves

- it also helps us to refocus our brain.

With self-love, you can:

- Create ways to learn to believe and trust in yourself
- Gain confidence
- Feel better about yourself and accept your flaws.

There is no one size fits all when it comes to what you can do to self-love and self-care for yourself. For you, having a cup of hot chocolate at the end of the day does the trick but for some people, meditating in the morning is a form of self-care and self-love.

What is important is that you spend some time to discover what floats your boat, what makes you happy and what makes your day a little better. This is not a cure all for anxiety or depression, but it is something that can help with these symptoms, it can help reduce stress and so much more.

Quick Advice

While we must practice self-care and self-love, in an ideal world we can do this every day. But since this is the world we live in and there's always bad days and good days, there is no way you can and will have time to love and indulge in yourself every single day. What we will see in this book is how we can make efforts to incorporate these things seamlessly in our daily life (it's as simple as throwing out all those women magazines that focus too much on body Image). This book will explore the many ways that you can practice self-love routines that help you grow exponentially, increase your self-esteem and decrease your stress levels. It will also focus on spending time with yourself (and not about rewarding yourself all the time).

Reading this book itself is already practicing self-care and you're welcome!

SELF-LOVE IN ACTION – WHAT IT LOOKS AND FEELS LIKE

Self-love, when applied in our daily lives brings about feelings of inner peace, joy and harmony. It is a feeling of contentment of knowing you are meant to be here at this time, you are happy and grateful for what life has to offer you right now and you are also happy and blessed in the body you are in.

Whenever we feel like we are not being good enough or you have negative thoughts permeating into your mind, this is how self-love comes in and becomes your trusted ally in combating your negative energy. It is the antidote to your inner critic and the chicken soup for your critical thoughts.

Self-love is not what we have or do not have but instead, it is something that we do. It is a very, it is an action and it is a practice that you constantly need to work on and just like anything we can to be good at, we need to practice self-love even when we feel good about ourselves or things are going great. Self-love is not just on a bad day.

It is about how we are with ourselves in every moment. It is not about what we feel or what we think but mainly about how we respond. This is because everyone has a story and we all have things that we wish were different about ourselves it is only human. We cannot always control what happens around us, but we can control our reaction to these things. Self-love is about moving and shifting away from identifying with the critical side of things and step into a more loving and stronger role that transforms our pain into purpose.

RECOGNIZING OUR PATTERNS

Self-love aims to alter our destructive patterns, from thinking critically of ourselves we alter our minds to be more compassionate. Rather than condemning our perceived faults, we listen and learn with an open heart. It is not about fighting our thoughts or even denying them. Through self-love, we will also learn to identify our destructive thoughts and we will let them take us down. Instead, we will become more adept at finding that part of us that is unconditionally loving. We will also grow to respond to our inner judgments with the same tenderness as you would be giving to a 10-year-old when they tell you all the reasons they don't feel enough.

We all need to be unconditionally loving to ourselves and as adults, this is our responsibility and not the responsibility of our spouses or friends, family and partners. It's our job, an internal job of self-love. We are the ones who are responsible for our own reality, and it is not the job of someone else to fix it because it is ours to honor.

As you journey through this book and learn the ways to apply self-love, it is normal to keep asking yourself if it works. However, it is all about practicing it, being consistent and being grounded in it. It naturally becomes a habit.

REAL-WORLD SCENARIOS

Not sure how self-love looks or if it important to work on? Let's look at these scenarios:

- You are going in for an audition as a newscaster. You have spent the last several weeks changing your appearance, buying new clothes trying to tell yourself that if you did it, you'd fit into the ideal candidate of a newscaster. As time gets closer to the date of the audition, you start beating yourself up telling yourself that you are not going to get the job because you are not as experienced or not as pretty or you do not fit the ideal description. You feel you will not be able to handle the emotional reaction if you get rejected. Because of these thoughts, you do not attend the audition, or you went for the audition, got rejected and felt totally depressed thinking you are a failure. You are unable to pick yourself up and dust yourself and move on.

- You are going in for an audition as a newscaster. You asked yourself several times if this is the right job for you and you've heard the judgments come up in your mind based on all the challenges that this job could bring. You think about your journey to get here, your experiences, your presence and you write down the pros and cons of the job as well as your experience and passion for it. You transform your daily experiences by applying care and you realize that there are many ways this audition could go, but you will give it your best shot and not hold yourself on such rigidity. You are nervous walking into the interview, but you release all your inhibitions, show up and do the audition. You are aware that whether you get the job or not has got nothing to do with your worth and value and the worst that could happen is that you may not get the job. But if it comes to that, you tell that you will deal with disappointment, but you'll try again.

These are two similar scenarios but with two different outcomes. The question is not whether you are good enough or attractive enough or worth enough. It is whether or not you believe that you can do it. It is the path of believing that begins with practicing self-love.

What Self-Love Isn't

We often seek love from others without realizing that the first place to look for love is within ourselves. Self-love is not a new terminology and the concept itself has gained popularity over the years and plenty of people can recognize the relationship between mental wellness and the importance of self-love. People are also questioning the circumstances in which we may even require love from ourselves, much more than we need from the people around us. During this assessment of self-love, oftentimes criticism also arises and often, it is based on misunderstandings.

Here are some elements that self-love most definitely is not:

- Entitlement

Having a sense of entitlement means that a person believes that they are unconditionally owed something without even contributing efforts, merit or even context. This should not be confused with the idea of recognizing your worth. Based on your perspective of deservingness and humility, you could also find it difficult to assert your self-love worth. In this, it is also helpful to consider fundamental human needs. One could argue that care, compassion as well as acceptance are as important as food, shelter and water. To recognize your worth and be in need of self-love overall is not that hard to seek. Remember that self-love is not about why you deserve a billion dollars or a college scholarship, it is not something else or exclusive, but it is an aspect of humanity.

- Selfishness

When we focus on self-love, we are not opening a path towards self-obsession. While self-love is a reflection process of turning our energy inward, the benefits of this process are not selfishness. In order to care for the people around us effectively, we must first care for ourselves. Think about the emergency oxygen masks in-plane safety announcements the flight attendant tells us to put our own oxygen mask first before assisting others, including infants. This is not being selfish. We also would not dare to tell someone who practices this as being selfish. In other words, you cannot pour from an empty cup. Self-love is not egocentric, rather it helps you so you can help others.

- Narcissism

Narcissism is superficial and vain. In the process of self-love, it is quite the opposite. When you practice self-love, you go beyond the surface and it is not all unicorns and rainbows along the way. You recognize your worth and your needs, you are also aware of the goals you have and the objectives you want to achieve. Self-love needs the courage to distinguish your challenges and your weaknesses as well as the obstacles that come your way and how it is up to you to change your perspective to face these challenges. Self-love is about empathy, humility as well as concern for yourself and others around you.

- Sinful

A sin is an act out of choice that is not only violent but also inappropriate. Self-love again is not sinful and it is the opposite of sin. Self-love does not tell you to go beyond or against your principles, beliefs and moral grounding. It is the enlightened journey to care for yourself in order to give off that domino effect of compassion and care for those around you.

Because of all the varying ideas of what self-love truly is, individuals from time to time may see self-love as being against their beliefs and values. Everyone is unique and their interpretation of scripture may change as well. If you are finding it hard to differentiate self-love from sinning, it could be helpful for you to do a little reflection and research on whether or not a conflict could truly exist.

No matter what your belief system or faith may be, it is worth considering the commonalities in various world religions especially where self-love is concerned. More often than not, being a person is a strong moral compass includes forgiveness, benevolence as well as personal growth, all of which aligns well with the practice of self-love.

- An excuse

Self-love is an engaged and active process that brings about a wealth of positive benefits. The journey to get there is not entirely an easy one. The major and most crucial component of self-love is recognizing our limits, worth and needs and asserting interpersonal and intrapersonal boundaries to uphold our values. This is an all-encompassing process that may include massages, mental health days as well as indulging in our favorite home-cooked meals but it is always not exploitation of all things good. Perpetually pampering we can also lead to negligence and it is a big distinction of what self-love is. The entire and full process of self-love includes bad days, ugly days and of course good days.

Self-love is also about looking at our areas of growth and recognizing the negative things we need to work on, and it is also about creating a plan of attack and bravely tackling it. To the untrained eye, self-love could appear as an excuse however it is imperative that you do not use self-love as a pass or reason to escape accountability and responsibility.

Takeaway

The process of self-love starts with the simple task of being able to appreciate yourself for the way you are. It is imperative to be considerate and kind towards yourself, but you should know that self-love is more than just a sentiment. Beyond your ability to tend to yourself, you must also remember that self-love is an intentional practice and one that has to be learned and cultivated. It gives you the ability to see yourself completely as yourself and to also recognize and value your weaknesses, strengths, challenges and triumphs.

When there is an emphasis on the self, this journey will lead to a more independent one. It is extremely helpful to unite with the people who are on a similar path as you so you can continue to uplift each other but then again, self-growth is dependent on personal effort and it is a subjective experience one that works differently for different people. You need to honor your individuality as you continue working on self-love.

Why Is Self-Love Important?

A question many people ask but so few understand. Self-love is important because let's face it you may be lacking in love, and it is also dangerous to leave the obligation of your emotional state (whether it's positive or negative) to the responsibility of someone else.

Loving yourself doesn't necessarily mean developing a shell and blocking out the world. It also does not mean having an ego so enormous that it prevents you from letting someone else into your life. It also does not mean forgoing relationships because only you enjoy being alone with yourself. And while it is tempting, it also does not mean that you can only live with pets for the rest of your life.

Self-love, while it is about being comfortable with the person you are, it also means recognizing someone else who does not know what unconditional love us and is having a hard time giving and receiving it. It also means recognizing that as human beings, we will hurt and that we could also pass this hurt along to other people. Through the process of self-love, we can understand how you can turn this attention of waiting and wanting someone to make you whole to realize that you actually have that ability inside you to do it.

<u>Points to keep in mind</u>

- No two people are exactly alike. Even though we may share many genetic predispositions, each individual has their own set of fingerprints by which they are known. Even identical twins are completely unique.

- Each person has inside them the capability of evolving in their own journey. Beating yourself up because you did or didn't do something within a particular timeframe is useless. People who have overcome obstacles have made a lot of mistakes along the way. Embrace your journey and know that you are learning as you go along.

- Each moment is a gift and never wait to apply self-care and self-love. There will always be plenty of things to worry about and to do and in the midst of it all, each moment gives you a chance to just stop and realize your importance in this world.

- Nobody is perfect. Every one of us has our own strengths and opportunities. Do not sell yourself short or even compare yourself with someone else. You are already setting yourself up on a losing proposition. Own what is positive about yourself and enjoy these things because you do not know how long you get to enjoy them.

- Practice self-talk that builds you up and not one that tears you down. Instead of focusing on how terrible things are, talk about how you can come out of these terrible things.

This journey as explained previously takes time and does not happen overnight. It is always good to take small steps to reach your goal.

A Journey Towards Self-Love In 10 Steps

In this book, we will cover 10 steps towards the journey of self-love. These steps are:

Step 1- Create Your self-love ritual

Step 2- Build Your precious community

Step 3- Create "What's Working for Me" list

Step 4- Know that your body is a loving vessel.

Step 5- Clean out Your closet

Step 6- Self Compassion

Step 7- Explore your spirituality

Step 8- Focus on Something You are Good

Step 9- Find your happy place

Step 10- Enhancing your Emotional Intelligence

CHAPTER 1

CREATE YOUR SELF-LOVE RITUAL

Among the best ways to begin our self-love journey is through using and practicing self-love rituals. These rituals are an easy and enjoyable way to instill and build self-love practices and behaviors until it comes naturally to you. Self-help rituals help you connect, heal and come back into the alignment of your mind and body with all that makes you, you.

WHAT ARE SELF-LOVE RITUALS AND WHY ARE THEY SO POWERFUL?

A self-love ritual involves a series of actions or steps which are a type of behavior that is invariably and regularly followed to achieve a certain state of mind or behavior. In a self-love ritual, we all want to behave or practice certain things regularly, consistently, invariably and habitually until it becomes second nature to us, and it becomes part of our lifestyle.

This is why developing rituals are powerful because rather than us leaving it to the randomness of life or maybe by chance or how we feel on that day, we create rituals and incorporate them into our lives, and this makes us more likely to do them. Once you have created the ritual that works for you, you don't have to think about it, it becomes part of your system and it is internalized and practiced easily.

Rituals have the ability to supercharge your habits and, in many ways, they are better than habits. You can encompass more than one habit in a ritual because rituals focus on specific ways or methods of doing something. Rituals are very focused.

Apart from creating a habit, the other amazing thing about rituals is that when we follow these set of ways of doing something, we naturally create space and time in our lives for it. Doing these rituals again and again sends a message to the brain telling it that this is extremely important for you and to you. You Matter.

If you are thinking that you need to layout a whole bunch of things, light candles and sing Kumbaya to do these rituals, well you will be surprised to know that it is nothing out of the ordinary. Your ritual can be anything you are comfortable with and by virtue of you thinking of it as a ritual and labeling it as such already makes it a thing. It is different from the mundane things you do every day and it is something special. It becomes that you look forward to it.

Your "You" Time-Creating A Ritual Just For You

Most of what you would consider a self-love ritual is things that you consider YOU time. It is simply just time you put aside to focus on yourself and scheduling this YOU time is a key habit in self-love. Having these rituals on and off are a great way for you to instill self-love How you do it, for how long, what it is and how easy it is for you entirely depends on you, your lifestyle, your situation and your preference. We have a few ideas for you but first, let us look at how you can create your very own self-love ritual.

Step 1: Set aside time to think of what you like

You just need to set aside or clock in an hour this week to really think about you and what you like to do, what floats your boat.

Write them down in a journal or even on your phone. Here's a quick table to help you with your thinking:

If I had 5 minutes to myself, I would like to:	If I had 10 minutes to myself, I would like to:	If I had 30 minutes to myself, I would like to:
- Put on a face mask	- Meditate	- Do a HIIT workout
- Have a cup of hot tea	- Sing out loud in the shower	- Practice some yoga
- Sit quietly in my car	- Do a few stretches	- Have a bowl of ramen
- Listen to some music	- Sleep	- Have a nice hot bath
- Massage my feet	- Call a friend to talk	- Bake a cake
- Eat an Ice cream	- Have a glass of wine	- Go for a run

Step 2: Create a plan

This step is more of creating or allocating the time you have on a weekly or daily basis to do the above. How much time you can spend on your ritual and how often you can do this ritual.

How long can I spend on my ritual?	How often can I do this?	What could prevent me from completing my ritual
- I can only spend 5 minutes a day	- Maybe only 3 times a week	- I could end up working late
		- My child would need my attention
		- The rain would disrupt the metro service

Step 3: Making it a Habit

After you have crafted some idea of a ritual, you now have a written or visible plan that you can see and one that you can envision yourself doing. The point of doing the two steps above is so you can already start to think about the time you have, how often you can spend on a self-love ritual and what are the kinds of things you can do.

There is no fixed guide to your self-love ritual. You do not have to stick to one ritual throughout your lifetime and you absolutely do not have to do a self-love ritual every day. You just need to have some rituals to use when you feel like you need to realign your thoughts and needs with your body and mind. One day it could be listening to your favorite music as you shower since you only have 5 minutes and the next week, it could be you completing a 30-minute bodyweight workout. You can do a self-love ritual every day if you have the time or only when you feel you need to.

If you're overwhelmed at the thought of adding one more thing to your to-do list, here are some ideas to get you feeling motivated, in control and in charge of your time.

If you only have 5 minutes	If you only have 10 minutes	If you have 30 minutes	For the Busy Mom
• Listen to your favorite songs on Spotify	• Take a hot shower	• Get a workout in	• Practice controlled breathing
• Give yourself a 5-minute back massage	• Light some candles and meditate	• Do some yoga, Zumba or dance workout	• Repeat positive affirmations
• Have an apple away from the computer, work desk	• Light some aromatherapy candles and just sit or lie down in its stillness and scent	• Read a book	• Go for a quick wash and blow
• Sit in silence for 5 minutes	• Dance to your favorite songs	• Go for a run	• Go for a walk with or without your kids
• Power nap	• Eat your favorite dessert	• Sit in a coffee shop, have something to drink and just be	• Eat that ice cream!
• Make a cup of coffee and enjoy the scent, aroma and flavor	• Read a chapter of a book	• Unplug from all devices and stay in stillness	• Have a snack

If you only have 5 minutes	If you only have 10 minutes	If you have 30 minutes	For the Busy Mom
• Hug your dog	• Put a face mask on	• Sleep	• Workout with your kids- use them as weights
• Hug your partner or spouse for 5 minutes	• Roller jade your face	• Watch an episode of your favorite show on Netflix	• Take time between your kids' schedules to just close your eyes and breath
• Talk yourself and give positive affirmations	• Have some ice cream	• Fix yourself a hot cup of chocolate	• Wake up early before the household wakes up and meditate
• Scream into a pillow	• Buy yourself new bedsheets or soft pillows	• Call a friend and talk	• Have a bar of chocolate
• Soak your feet	• Do 10 minutes of skipping	• Listen to an uplifting podcast	• Cuddle with your kids
	• Pray	• Take a walk	• Watch funny videos
		• Give yourself an oil facial massage	• Scream into a pillow
		• Gua sha	
		• Pleasure yourself	

BENEFITS OF SELF-LOVE RITUALS

As humans, we are creatures of habit and along the way of life, we pick up both good and bad habits. As Aristotle once said, "We are what we repeatedly do' and in that sense, we need habits and rituals because it helps in guiding us and giving our lives a sense of rhythm that we can dance to. It also helps us see what habits need to be removed, what needs to stay and what can be improved. We also use rituals as a way of reminders of the things we care about whether it is getting to work on time, of making our beds or making sure our presentation has proper labels or ensuring our kids have little notes to say how much we love them. These are all rituals. It keeps us all close to our purpose and it also defines our values.

You may not have realized this, but you probably already have your own rituals without even thinking of them. It could be a Sunday night dinner with the family or visiting your parents for Friday dinner. It could also be those positive affirmations we repeat to ourselves in the shower or the prayers we say at night. Washing your face is a ritual and so is having that mandatory cup of coffee. The difference between your daily routine and rituals and the self-love ritual is that the self-love ritual brings you back to you. It is to honor you and to focus only on you and nobody else.

Here are some benefits of self-love rituals if you want to explore them:

1. Self-love rituals bring your awareness and attention to the present moment

There are some rituals which you can do daily and there are some that are more of a once-a-week thing. However often you do them, these rituals give a slight or long pause in the day to enable you to focus on your intentions. They are things you do that are fast which are things that are important to you such as kissing your partner goodbye before leaving the house. Or it is a long ritual such as spending some time with nature before heading back to work or home. The attention we give to these rituals enables us to embrace it with our energy.

2. It is an action step that gets you focused on action steps

The self-love ritual creates a domino effect that sets in motion your entire day. Rituals like taking 5 minutes every morning to stretch or to meditate or even 10 minutes at night to watch your favorite Netflix episode are all considered action steps. We engage in them for conscious purposes which is to live the way we want. Taking time to do these rituals is important and it makes us come closer to our goals. If our intention, for example, is to ensure that we connect with our family members by wishing them good morning, then we are more likely to follow up with that action.

3. Self-Love Rituals Brings people together

You cannot pour from an empty cup. One you do yourself right, you pass on this positivity to the next person and they feel one step closer to you. Oftentimes, self-love rituals are done in solo, but it could also be in the form of support groups, picnics with friends even church services. Whatever we do, it defines who we are, and it also contributes to our identity. We need others to reflect back to us on who we are and what our significance is. This will help us going forward living and acting more closely with those ideals.

4. Reminds us to appreciate

Self-love rituals in a way is a method of giving thanks and being grateful for the life we lead no matter how hectic or stressful it can be. It is us taking time to be mindfully thankful for the things in our lives and it helps us to not take things for granted. When we are more grateful, it also improves our outlook on life, and we radiate positivity to the people around us. It also gives us more energy to create more goodness around us. Gratitude also gives us better tolerance and patience for hard times. It is always a win-win situation.

5. Creates meaning

Last but certainly not least, self-love rituals give us meanings. Things that bring meaning have more value to us and brings out more attention and energy. Rituals help us understand and remember why we do what we do. It helps us embody what is most important to us, the commitments we have made, the beliefs we have embraces as well as to celebrate our accomplishments, honor the community we are in and also to show appreciation for all the things around us. Commitment to a ritual is hard to sustain.

Rituals fill us with a robustness that encourages us to go on and continue on our paths. Rituals make us stronger.

Chapter 2

BUILD YOUR PRECIOUS COMMUNITY

Being part of an encouraging and positive community has a strong and tangible impact on people, their self-awareness as well as their fulfillment. When an individual experiences positivity with their communities, it allows them to feel more connected to their environment and the people in it. This connection also becomes a support system for them when they require sympathy as well as encouragement.

Strong feelings of connection to a group will also prevent any mental illnesses that could crop up due to alienation, anxiety and depression. People who experience positive community experiences cultivate a sense of belonging and can also openly express themselves without feeling like they are being judged. Communities that excel at encouraging their members to voice their opinions about their ideas and thoughts often create a membership that considers their positions from a deeper perspective.

SURROUNDING YOURSELF WITH LOVE AND SUPPORT

Every relationship we have with someone is different: some are good, some are bad but all of it matters. You form relationships and connections with people all around you, from the butcher down the street where you get your meats to the supermarket cashier you say hello to when you get your groceries. All of these are the connections you make. These people remember you and your smile and would probably make an extra effort when they see you the next time.

Your relationship with the butcher or the cashier may be quite different from the ones you have with people closer to your social circle but all of them play an important role in the community that you are a part of.

The more connections you have, the better but as you get to know your community better, strife to have more meaningful relationships with the people around you. You never know when they would come in handy. Whether they are schoolteachers, government officials, the gardener, the daycare center operator, homeless people they are all friendships waiting to be explored and this can pay off for you in plenty of ways than you expected.

YOU AT THE CENTER

You are at the center of your ecosystem and each system represents your relationship with another person. Just like the spokes of a wheel that holds the wheel together, your relationship with the people around you is what helps move you and there is enough room for everyone to create their own wheel of strong relationships.

You have to take time to set up and sustain your relationships because it is a two-way street you need to put in as much work in the relationship as much as the other person is. If you wait for someone else to establish this relationship, you may end up spending a lot of time waiting.

It does not make sense to form connections and relationships just for people to work for you and it will not work in the long run because people will feel used. Build your connections with integrity and because you genuinely want to not just because of what someone has to offer you. Build relationships based on a common goal.

How to build relationships?

Here are some tips to help you form relationships with your community. Especially if you are new to a place, or have been having trouble meeting people, this could be a good refresher for you:

- **Start small**: building relationships one at a time is a good way to start. You can begin by joining interest groups (find them online or on Facebook) or you can just start talking to the folks in your building. You can also become part of online communities and join meetups when they have one.

- **Be friendly**: it may seem self-evident but a smile or a friendly word can definitely make someone's day. You can find something in common with the people around you because, at the end of the day, everyone is just like you, they want to have close connections as well.

- **Ask questions**: people love talking about themselves and about what they like and what they think of certain things. When you ask someone about themselves, take the time to listen and this can definitely make them a fast friend.

- **Tell people about yourself**: as much as you ask questions, you must also be open to telling people about yourself too because this is a way to establish trust. People will be able to trust you unless you are willing to trust them with the information about yourself.

- **Go places and try things:** If you want to make friends, you need to go out and be at places where people are. Join conferences, events, picnics, fundraisers and parties, go where you will meet people.

- **Accept people for the way they are:** You do not have to agree with everything all the time in order to form a relationship. But on the other hand, people do not like to be judged. Just because they do things differently than you is not wrong.

- **Assume that people want to form a relationship too:** not everyone will be ready and open to form relationships but even the crabbiest of people is a lonely soul hoping to form a bond with someone

- **Overcome your fear of rejection:** We all have a fear of rejection at varying degrees and the only way to get rid of it is to get over it. If you want to form connections and relationships that are beneficial for you, put yourself out there. If you get rejected, it is their loss.

- **Be persistent:** most people are shy and not everyone is extroverted enough to make friends and start talking. It will also take a while to win trust. That said, form connections anyway and trust will come when you stick with it

- **Get involved:** Many of us are looking for opportunities to meet other people who share the common goals as us. So, you get involved and also invite people to get involved with you. They will be flattered that you invited them to join.

- **Enjoy the company:** Having people around, talking and making friends is actually a really nice thing to do. If you genuinely enjoy having people around you, then others will feel this energy and be attracted to your attitude. They will also be more likely to be around you.

COMMUNITIES WOMEN CAN BE PART OF

It is increasingly difficult for women to find safe havens in today's world and many of us are moving to virtual space to find and meet connections, share our stories and pitch our businesses. Online communities centered around women create protected spaces where women can freely express themselves and let their guard down.

If you are looking for communities that you can be a part of in your neighborhood, but you don't know where to start looking, then the internet is your main source.

One of the first places that you can start looking at is Facebook. Just go to the group section and type in the groups you want to be part of whether hobbies and interests, arts and crafts, fitness, relationships and identity, networking and many more.

There are many women out there who have stories from all walks of life and from various things like their struggles, their transitions, shame, vulnerability and all of them want to voice out.

Another online community that you can also be part of is Women Making Waves, for women "seeking inspiration and connection" to share stories about their struggles and vulnerabilities. You can also seek out communities that are on the same path as you, self-love and join these communities to talk about topics that may otherwise be taboo.

If online communities do not work for you, then try looking at associations such as the Young Women's Christian Association, church associations, fitness clubs and so much more. Again, doing a quick internet search will get you what you want and where to find these associations and clubs.

You can also join blogs to voice out issues regarding identity, safe female traveling, self-love, self-care and so much more. Not only will you get to share your stories, but there is a real sense of bonding and a feeling of sisterhood.

The thing about connecting online is that it expands beyond technological interfaces and these communities plan real-life events, meetups, seminars, coaching sessions and so much more. The online communities bring digital space into the real world. The convergence of the internet to the real world is important because although internet communities enable people to connect beyond geographical borders, nothing replaces a face-to-face meeting.

The human brain is designed to look at a person, see their faces, their eyes, feel their skin and see all those micro-facial movements. Via the online avenue, you extract a very thin thread of what our brains can perceive, making the level of connectivity and nurturance different. It can never be replacing the human-to-human connection in real life.

Joining women's groups, be it for hobbies, for fitness, for business it is all a form of support where you can thrive without feeling like you are facing these challenges alone. Being able to talk about your struggles with other women is a breathtaking journey and it takes a load off your shoulders because you know someone else who is going through what you are going through and both of you can support each other, keep each other accountable and motivated.

CHAPTER 3

CREATE "WHAT'S WORKING FOR ME" LIST

In this chapter, we explore what is called a self-acceptance journey. This is in no way a MUST follow approach for you. It is just to give you an idea of what your journey could be and should be and may a little about what self-acceptance is.

We all have our own unique and individual journey and our paths in life are all different. All of us have different timelines in how we encounter this little thing called life and it all has its own ups and downs, twists and turns, flat roads and high roads. It smoothes for some people, bumpy for some but no matter how easy or difficult this journey is for you, it made you, YOU.

JOURNEY TOWARDS SELF-ACCEPTANCE

Self-acceptance is not something that you learn in school. For many of us, it is something that we learn gradually, or could it also be achieved by claiming it?

The topic of self-acceptance is a funny one. It is a much talked about a subject that often comes up in personal development groups, mentoring groups, spiritual conversations, and even self-discovery workshops. It's everywhere.

The common wisdom is that we should all accept who we are. Isn't that what everyone keeps saying? 'Be yourself', 'Be Authentic', 'Why do you want to be me when you can be you?'. While all of these are uplifting and motivating, we all know that the path to self-acceptance is not an easy one. It is harder to accept our own shortcomings than it is to accept other people's.

So, what is self-acceptance?

- It is something you make a choice about

On a mental level, you can practice some form of self-acceptance by seeking ways to gain it such as using positive affirmations or even going through therapy. But in all honesty, it is through life itself that you will encounter situations that challenge your thought process and it will either help to strengthen your self-acceptance or you may end up questioning your beliefs about yourself.

- It is not a final destination or an end goal

There will always be times that you will question yourself and there will also be times that you will accept your strengths and weaknesses. For the entire time that you are alive on earth, you will go through changes in your self-acceptance journey. There is no final destination for self-acceptance my friend, except death. An attitude shift makes a huge difference but no amount of theory of self-acceptance will help without the actual practice.

- Life shows you ways towards self-acceptance

Life has a funny way of showing us how to become accepted through the challenges it throws you. You must see these challenges as an opportunity to develop your character and to test your limits. This is your potential and your opportunity for personal growth. There will be times when you want to hide yourself and your flaws. There will be times that you will feel like you are not enough, or you just want to be like all the girls around you. You may also be suffering from things like PCOS or even back acne or you've had a traumatic childhood, and nobody knows about these things.

But with life, there will come a day that you will feel the need to talk about these things and if this happens, do not ignore the signs. Life (or divine intervention, God) will give you the motivation to accept these things about you and you can choose to talk about it if you want, or you can choose to just accept them and realize that they make you, YOU. The more you accept it, the more you'd be okay with it and you may start sharing your stories with people and be more open than before. When you do this, you'd be surprised how many people have gone through what you have or are struggling through these things on their own too.

- It makes you open about your challenges and circumstances

When you feel self-acceptance for yourself, you are more open about your challenges in life. You talk about it, you share it and, in many ways, you enlighten people about certain issues and you also empower others who have been through what you have or are going through the same thing you have. There will come a day when you can openly discuss your issues. It will take most of your life to get here but you can look back and appreciate the journey to get where you are now.

Self-acceptance isn't something that happens overnight, but it's something that you can consciously choose to use and adapt as your attitude when you practice mindfulness as well. As human beings, life will continue test is in many different situations and each of these experiences will reveal just how much self-acceptance we really have for ourselves.

As you go through these experiences and as time goes by, you can be sure that you will reach a greater and higher level of self-acceptance. The decision towards journeying in self-acceptance begins with you and it starts with you telling yourself "I love and accept myself."

THE "WHAT'S WORKING FOR ME" LIST

Where do you find inspiring ideas for your projects? Did you find it sitting at your work desk? Some people find inspiring ideas during their runs or hikes up in the mountains. Some people find inspiration while playing basketball with their kids. Some people find an idea when they are putting the finishing touches on a birthday cake they were icing.

All of these things are ways people get inspired, in other words, this is their 'What's Working for Me' list. It doesn't necessarily have to be something that they are doing. For instance, some people are inspired or motivated when they are lying down in the afternoons doing nothing except looking up at the sky and eating a sundae cone. It helped them accept silence as well as listen for inspiration.

In the journey of self-love and self-acceptance, it is about creating a 'Love-To-Do' list rather than a To-Do list and once you do this, you'd be more satisfied with both life and work. This is not to say that you should throw away your to-do lists and screw work. Nope! We are not asking or suggesting that you do this. You'd still have to create that presentation; you'd still need to sign up for that PTA meeting and you'd still need to put in that 30 minutes of cardio, it is life and it needs to go on. The objective of the 'What's Working for Me' list is to help you approach work tasks without a sense of dread. It is about creating something that is energizing and enjoyable past every task and that you have a motivation that feels pretty much bottomless. It is about channeling your positivity from your What's Working for Me into the daily, mundane activities of life.

For instance, if you woke up this morning and went on a 2-mile run, you got back home, put on a fresh shirt only to have coffee spill on your suede heels. You say, 'Thank goodness, not on my shirt'. You wipe off any stains and get on with life without a sense of dread. Or it could also be reading this book. It may not have been the apex of your self-love desire when you woke up this morning. But its something you did anyway because you channeled that early-morning positivity generated from your run session into something productive.

The problem with what all of us are facing is that our society and culture has this huge amount of lazy shame. We can't live with ourselves if we are not producing something. So, when the weekend comes, you need to be guilt-free. Do not have the need to feel like you need to be busy for the sake of being busy. Of course, not every weekend is going to be a weekend that isn't busy. But when you do have weekends that have no agenda, no workshops, no soccer practice, relax and recharge either on your own by doing your self-love ritual or with your family. When you settle into your keyboard come Monday morning, you'd be less pulled back into the Monday blues that most people seem to have. You do not have this because you know that you can also do a little of what you love or a little of your self-love ritual on a workday too.

SCHEDULING YOUR 'WHAT WORKS FOR ME' LIST TAKES DISCIPLINE

You do not need to quit your job in order to have more time to do what fulfills you. Your boss or your team probably will not approve if you left early every Thursday afternoon to go hiking. It is true that work is still work and yes, it is all not fun, but this has nothing to do with work but rather how YOU build more 'love' into your to-do lists.

It is more about consciously and regularly doing more of the things that work for you and in this way, you cultivate more energy and inspiration to excel at what you do. Just like anything else, it takes discipline to cultivate it into a habit. Apart from your self-love rituals, you can also start reflecting on the activities that you have done that made you the most energized. You can start by listing the things you loved doing in your self-love rituals or anything else that you did over the average week.

If you feel like you couldn't find anything to be happy off, then take a step back and think about the things you used to do prior to having kids, or prior to getting married and what are those things you could try to do again. Once you have listed your 'What Works for Me' list, you can be a little bit more wistful. Write down the things that appeal to you even if you have never tried them, like bungee jumping or salsa dancing.

Next, this is where you need to commit. You can pick one or two of these items and realistically accomplish them in the next few weeks.

You can schedule as part of your ritual or you can make this a weekend thing. These are the appointments you make for yourself that lock into your calendar just like any other activity or event. You need to keep just as much as you need to not miss that conference call or meet that project deadline.

As an example, here is what your daily What-Works-For-Me list can look like:

- Meditate
- Journal
- Do some sprints
- Breakaway from the computer every 30 minutes for a round of pushups
- Read some fiction
- Listen to your favorite 90s songs
- Sing
- Play with the dog
- Reflect on the things I'm grateful for
- Look at the clouds

You do not always have to check off every single thing on this list every single day, but you can try to hit most of them whereas before writing your What Works for Me lists, these activities were just periodic pastimes. And for your week, you can schedule bigger activities that can't be done on a daily basis such as:

- Take a long hike in trails
- Practice basketball at least twice a week
- Go swimming with your girlfriends at least twice a week

- Play tennis doubles on a Saturday

- Help out at the soup kitchen

- Spend time with your parents

Commit to your rituals and to your love-to-do lists. Not many people do this which is why daily life is boring and mundane. Once you begin to do this, you will find out that you not only got better at keeping your work-related commitments, but they will also start feeling less like the chores they used to.

CREATING A POSITIVE AND GRATEFUL MINDSET

The Greek philosopher Epictetus said that 'The thing that upsets people is not so much what happens but what they think about what happens.' This was said sometime in the early first century, and it gives us all an indication of the origins of positive thinking.

Subsequent great philosophers, authors, CEOs, advocates, activists and leaders have spoken out about having positive thoughts and seeing positive value in every aspect of life because how we look at things defines the realities that we experience.

THE ORIGINS OF POSITIVE THINKING IN MODERN SOCIETY

In modern society, we can attribute the origins of positive thinking to two publications that came about after the Great Depression of 1929. These books are 'How to Win Friends and Influence People' in 1936 by Dale Carnegie as well as 'Think and Grow Rich' 1937 by Napoleon Hill, 'The Seven Habits of Highly Effective People' in 1989 by Stephen Covey, Eckhart Tolle's book, 'The Power of Now' published in 1997 and of course, there's also the now famous 'The Secret' by Rhonda Byrne which focuses on the law of attraction and positive thinking. The various authors have gone through various research, observation as well as life lessons to conclude several points discussed in their book. Napoleon Hill, for example, interviewed five hundred millionaires of his time- both men and women to identify the secrets of their success.

THE CONSOLIDATION OF POSITIVE THINKING
Norman Vincent Peale in 1952 published a very Christian version of positive thinking called 'The Power of Positive Thinking.' Peale is also the founder of the 'Guideposts' magazine. Earl Nightingale followed up with the first positive thinking audio recording called 'The Strangest Secret' in 1957. His point on positive thinking is 'We become what we think about.' The 'Psycho-cybernetics' in 1960 by Maxwell Maltz explains how our mind works and how we can practice ourselves to use positive thinking and what kind of exercises we can do to train the mind.

All these books have had an impact on our culture and lifestyle at various points in history, and they have helped numerous people to achieve success over time. There are even learning institutions today that teach principles found in these books as well as seminars and conferences and get people who want to learn, develop a more positive outlook on life. ACTION coaches use several of these elements in their coaching.

POSITIVE THINKING IN THE TWENTY-FIRST CENTURY
Today, positive thinking is applied in many different fields from business to sales, marketing to advertising, health, sports, education, motivation, inspiration, national allegiance, psychology as well as self-image. Many of the twenty-first-century authors apply positive thinking in various areas. Some of these famous ones are:

• **Anthony Robbins'** seminar and speeches using the knowledge of psychology and positive thinking. Robbins' is a motivational speaker and advisor to many world leaders and has helped ordinary people to achieve success or lead a more positive and fulfilling life.

• **Steven Covey** is the author of 'The 7 Habits of Highly Effective People', and his points are regularly quoting in businesses and personal development. These seven habits can be used above and beyond the business realm, applying it to almost anything in life.

• **Louise Hay** is the author of 'You can Heal Your Life' and several other motivational and self-improvement books. She promotes the use of self-healing to use the power of our thoughts to enhance our lives.

• **Wayne W. Dyer** employs the teaching of Tao Te Ching of 'Change your thoughts, change your life' which directly influences use to lead and live a more balanced and fulfilling lifestyle. Dyer is the author of 'The Power of Intention.'

Why Positive Thinking is important for one to truly live an abundant and productive life:

How many times have you failed at something and someone a friend, teacher, classmate, parent or partner tells you not to give up and focus on the positive?

Sometimes you think that it's easier said than done. The truth is, focusing our mind on being and thinking positively is fairly straightforward it is all about controlling your thoughts because of the understanding that a positive attitude leads to a fruitful, and happy life is already a high motivation to change. Having a positive outlook on life will enable you to cope much more easily with the affairs of everyday life from the moment you wake up until the time you go to sleep. A positive outlook gives you an optimistic approach and makes you worry less and think fewer negative thoughts. It will enable you to experience the silver lining in the darkest of situations. A positive mind is a state of mind that is worth developing because everyone can benefit from it and who knows where it will take you.

A positive attitude is noticeable in the following ways:

- Positive thinking.
- Constructive thinking.
- Creative thinking.
- Optimism.
- Drive and energy to do things, accomplish goals.
- An attitude of happiness.

A positive mindset can help you in many ways:

- Expecting success as failure is not an option
- The feeling of inspiration in everything you do
- Gives you the strength to keep going and not give up
- Helps you overcome obstacles you face
- Gives you the ability to look at failures, mistakes and problems as a blessing in disguise.

- Keeps you believing in yourself, your abilities and your talent

- Radiate self-esteem and confidence

- You look for solutions instead of dwelling on problems, you seek opportunities when it comes

Positive thinking is a game changer: you can change your whole life if you always look on the bright side of life instead of wallowing in self-pity and allowing yourself to think negatively. Positive thinking is infectious! It not only affects you but each individual around you, people want to be with you and make friends with you and hang out with you because you've got the drive and energy and positivity, making it so easy to be your friend. You will end up changing the lives of those around you, uplifting them and encouraging them to become the best version of themselves. Positivity is a strong emotion so if you are positive; you radiate positivity.

Even more benefits of a Positive Attitude:

- You achieve more of your goals easily

- You achieve success much rapidly

- You bring in more happiness in your life and those around you

- You have more energy to deal with everything life throws at you

- You have more faith in your abilities and have higher hopes for a brighter future

- You can inspire and motivate everyone around you

- You feel you encounter fewer obstacles and difficulties compared to other people

- You are much more respected and loved by all those around you

- Life smiles at you

The bottom line is, if you exhibit a negative attitude then you will only bring in more failure and more difficulties but if you radiate positivity, you are bound to be attracting good energy and success so the time is NOW to change the way you think and the way you react. Negative thoughts, behaviors and reactions do nobody any good. If you have tried to become positive in the past but you have failed, that means you have not tried enough.

CHAPTER 4

KNOW THAT YOUR BODY IS A LOVING VESSEL

A healthy individual is one who also has a clean bill of health. An individual with a clean bill of health is at a lower risk of contracting any deadly diseases which include cardiovascular problems, diabetes, arthritis and more. This is what you should strive to be. This is what loving your body is, it encompasses our mental health, emotional health and when it comes to physical health, it is just not about how good you look on the outside but how healthy you are on the inside.

Exercising regularly is something that you need to start doing as well when we talk about being mindful about our body if you're not already doing it. If you want to be healthy, exercise is going to have to be a part of that package. Not only does it help to regulate your body's blood flow, but it also helps to balance out and boost your metabolism, which goes without saying is needed for overall good health.

BEING MINDFUL OF OUR BODY

You Become a More Confident Person – Living a healthy lifestyle can do wonders for your mental and emotional health. You're going to start feeling really good about yourself because you like the results that you see in the mirror. A healthy diet and regular exercise will give your body that healthy look and glow that no cosmetic product or makeup would ever replicate because that glow is going to come from within. When you know that you look good, you start to feel good. You're energetic, you're clear-minded, you feel good, and you feel like you are ready to take on any challenge that is thrust at you. You become a more confident person.

You're Emotionally More Stable – Our emotional state is linked to good health because believe it or not, the mood is psychosomatic. When your body does not feel good, don't expect your mind to be feeling all that great either. Mood swings will become inevitable when a person experiences hormonal ups and downs as a result of poor health. If you find this hard to believe, just think of a time when you came down with a cold and felt fantastic emotionally. Almost certainly the answer is going to be never because it is impossible for us to feel any kind of happy when our bodies are not working as it should be. That is a perfect example of why good health is so important because being healthy will help your body boost its serotonin levels, which is also known as the happiness hormone. Yes, that hormone exists, why do you think you get that rush of happiness when you consume alcohol or ice cream for example? Because of the temporary spike in serotonin that you just received. What to feel that way all the time? Be healthy.

You Obviously Look Much Better – This goes without saying. Sugary drinks, fatty foods and junk food may give you a temporary glow and feeling of happiness, but that is only going to be short-lived. Why? Because eating too much junk like that is also going to cause you wrinkles and some extra, unwanted padding around your waist. Plus, it makes you feel lethargic and sluggish when you are carrying around more weight than you should, which is what a lot of obese people feel like all the time. Being healthy and consuming enough water, getting ample sleep and eating a nutritious diet packed with fruits, vegetables, grains, protein and more is the secret to keeping your skin looking healthy and your hair shiny. Not only are you going to find that you look better, but your movements are going to become a lot more energized, which will give you an overall look for appearing more attractive, youthful and supple.

You Become More Productive – Because we can't always rely on coffee and tobacco to give us that spike in energy that we need. It's only a momentary spike and it won't last long; wouldn't you prefer having that feeling all day long? A balanced lifestyle and good health are how you achieve that same spike in energy that coffee can give you each morning. Cut down that dependency of getting your energy sources from elsewhere and instead, just choose good health to feel that way all the time.

It Saves You Money – Being mindful of the way we eat and the money we spend on food will help you save money too. You make mindful choices of the food you eat as well as how fresh it is. Your meal prep, you regulate you're eating hours and you spend more time cooking healthy and nutritious food.

It Helps You Lose Weight & Keep It Off – Needless to say, this is probably one of the biggest physical benefits when it comes to eating nutritiously. It's not just about losing weight, equally important is keeping the weight off for good. This goes back to the earlier point about not restricting yourself to only certain food groups and being able to eat regularly so you don't feel like you're depriving yourself in any way. The happier you feel about your diet and being able to eat the foods you like is the difference that makes intermittent fasting more effective than anything else out there.

You Become More Confident & Disciplined – When you look good, you feel good, which is why people embark on a weight loss journey, isn't it? Because they want to look like the best version of themselves. Intermittent fasting can do that for you, and not only will you end up looking and feeling much healthier, when you see how good you look in the mirror your confidence level increases. Your confidence levels will also be aided by the fact that eating intuitively has taught you to be a more disciplined individual, focused on a goal, and being able to accomplish a goal that you set out to do. You need some level of discipline to succeed even with intermittent fasting because you need to be able to say no in the face of delicious food when it's not your time to eat. Sticking to a regular eating regimen helps to build your discipline levels, and when you can do that, your confidence levels are one again elevated because you can look back and say yes, I did it!

MINDFUL EATING

Practicing mindful eating is one of the best ways to improve your relationship with food. This technique helps you gain control over your food and eating habits and it also promotes sustainable weight loss, reduces binge eating and overall, you feel much better after eating. In this chapter, we will look into what constitutes mindful eating, what you need to know to get started as well as how it works.

Is it starving?

Mindful eating is not starving the body. It puts awareness on the menu wherever and whenever we eat. It makes us watchful of what we eat, and it also aims to transform our relationship with food by focusing on the why and the how of eating, encouraging more sustainable and holistic points of view. This also means that we have better opportunities to understand how food nourishes our bodies and what it can do to keep us healthy. It also helps us create a deeper appreciation for the work that goes into making a meal, savoring every mouthful and every ingredient.

Is it dieting?

Mindful eating, on its own, is not a diet. It does not have any kind of juice cleanses and neither does it require you to eliminate any kinds of food. It also does not tell you when to eat and there are no promises of quick fixes. Mindful eating is more a framework or guide that helps you eat better and make better choices with your food. It puts you as the decision- maker of all your choices, so you choose what to eat in order to meet your fitness and weight goals. Through mindful eating, you are not eating just as a means to an end, it is not about choosing foods based on our desired outcome. This is potentially self-defeating. All mindful eating does is to invite ourselves to be present while we cook and eat our foods, to savor without any guilt or judgment, anxiety or inner commentary.

This method is all about spending less time focused on what your weight would be or should be and all kinds of storylines about your weight and instead, embracing eating in a way that helps people find the weight that is right for them. Conventional dieting and eating plans cause too much stress around eating, even if the plan wasn't meant for it to be like that but because we put in heaps of pressure and intensity together with false expectations, most of the time we end up ruining our diet and eating plans. Through mindful eating, we want to eliminate the view of food being a reward or punishment. Because we view food like this, we end up saying we 'deserve' a huge slice of chocolate cake because we view it as a treat after going through 24 hours of fasting or going one week on a celery juice diet or even eating less than 500 calories for an entire month. People who want to lose weight are obsessed with wanting to be thin that we either end up undereating or suppressing our feelings of hunger or we just end up ignoring our signs of being full.

The issue begins when we internalize ideas built around dieting and we get sucked into the idea that suggests losing weight is as easy as ABC. This is when the emotions and pressures are heightened, and we end up sabotaging our diets and plans after a while. Mindful eating is all about encouraging us to let go of the traditional all-or-nothing mindset and refocus our energy on eating according to our natural body weight and not the one shown in magazines or on TV. There is no definite strategy with mindful eating neither is there any calorie-counting on extreme ends. All we are doing is simply trying to be aware.

The mind is calmer when we are more aware and when the mind is calm, we are less agitated or stressed and the less we eat emotionally. When we are more focused, we also increase our clarity to look at the choices we make on food and make better attempts to eat better food choices. When we are calmer and clearer, we are more content, more compassionate on our weight and ourselves and we also become more aware that sustainable weight loss is a journey. Mindful eating is about bringing mindfulness back to the dining table and this also means a kinder and gentler approach towards eating. It is not about changing our diets but more about changing our relationship and thoughts about food.

WHAT ARE MINDFUL EATING HABITS?

As we know, mindful eating is about tuning our bodies and making it more aware of the sensations we experience that precede the 'fullness recognition' in our brains and it also helps us to better gauge our satiation state versus our snacking state even without waiting for 20 minutes. It helps us reach a state of attention on our cravings, our experiences as well as our physical cues. We reset our ideas of what hunger and full cues are and eliminate cues that we learned as a child such as "You're not leaving the table until you finish your plate!" or "You can't be hungry!" or "Are you sure you need seconds?". Our conditioning with food starts early and we cascade these ideas down to our children to help them listen to their states of hunger and fullness, rather than just ignore them.

Essentially, mindful eating habits include:

- appreciating your food, eating slowly and without distraction
- Understanding and knowing true hunger and non-hunger triggers for eating
- engaging your senses through colors, smells, sounds, textures, and flavors
- learning to cope with guilt and anxiety about food
- eating to maintain overall health and well-being
- Seeing the effects food has on your feelings and figure
- listening to physical hunger cues and eating only until you're full

EXERCISE TO LOOK AND FEEL GOOD

It is a known fact that regular exercise can protect your health and help you maintain an ideal body weight in the long run. But did you know that it can also yield additional benefits for the busy professional such as yourself? Studies have shown that regular exercise releases hormones to boost your mood, reduce stress, helps you sleep better at night and if performed in the mornings, help-jump start your day. In the end, exercise helps to boost productivity at work and ultimately boost your business success. Here are some examples of how regular exercise can help not only improve your health but your professional career as well:

- Improve Your Women's Network

Apart from networking functions, the gym or any sports- related activity can provide an opportunity for meeting individuals that can help strengthen and widen your existing business relationships. A lot of individuals resort to taking up yoga and Pilates as one avenue to broaden their circle, but other sports also do offer the same kind of opportunities as well. Even the gym also provides the same possibility as you tend to meet people with the same interests and background as you do. Exercising with others allows you and them to open up their personalities and communicate in a way that is different than you would have if you met them in the boardroom.

- Goal Setting and Determination

When we start exercising or picking up a new sport, we will always set goals for us to achieve. The training we put into these goals such as completing a marathon can help business professionals use that same determination and drive to pursue the challenges and objectives they face in the boardroom.

- Increase Your Confidence

Every time you hit the gym and accomplish a goal that you set for that workout; you begin to feel good about yourself. This accomplishment by having a great workout can translate into every area of your life. This can improve your confidence level throughout the day and make you feel great about everything you set yourself to do.

- Improve Creativity and Thinking Skills

Studies have shown that exercise does not only benefit the physical part of your body but also the mental part as well. It was shown that cardio exercise not only improved the aerobic capacity of an individual but also improved their cognitive functions as well. The increase in blood flow to the brain has shown to be the main cause of this increase.

- Reduces Stress

When we are stressed, we notice that we are unable to act and think properly thus we end up making a lot of not so good decisions at work. Exercising helps us battle this stress by releasing endorphins when we induce our bodies through physical activities. These endorphins give us a natural high and allow us to approach our day with a calmer mindset.

- Increase Energy Levels

Working day in and day out can lead to an increase in fatigue levels to an individual. If you don't fight this, you end up going to work with low energy levels that will inhibit your performance at work. In a study done on the effect of exercise on an individual's energy levels, it was shown that performing exercises on a daily basis helped them to increase their energy levels by 20 percent while it also helped to reduce their fatigue levels by 65 percent.

Exercising regularly not only helps you to be awake throughout the entire day but it has shown through this study that it helps you get better sleep at night. We have gone through some of the important points why exercising is important both physically and mentally to a busy professional like yourself. But if you have not exercised before or had trouble getting into a routine, it can be a daunting task to reap those benefits that we spoke of earlier. So, the best thing we need to do to build a routine out of exercising is to plan ahead. Let's see the steps that we can take:

- Choosing Your Workout Times

Choose a time in the day that you feel you can focus on and complete your workouts. This can be either early in the day or at night. There can even be a case where you can sneak in an exercise during lunchtime if your office has a gym. But ultimately, you have to decide what is best for you and a routine that you will most likely stick to base on your busy schedule. Another important aspect to remember is to always listen to your body. If your body is still sore from your previous workout, or you are feeling tired sick, best to skip the workout for that day. But don't skip the next one when you feel better.

- Choosing an Exercise

Once you have set your goals and workout time, it is no time to choose what workout you want to do. This can be as simple as going for a jog or doing jump ropes. Swimming is another popular exercise that is easy and convenient to do. Joining a local gym and signing up for their aerobic or yoga classes also is a popular choice amongst busy professionals. How about hitting those weights? Don't know where to start? Get a personal trainer to help you through the basic and fundamental movements. The most important criteria to remember here is to select an activity whether it be a solo or a group activity that most likely you would stick to in the long run and become a part of your daily routine.

- Learn to Have Fun

The one final thing that you will need to know once you have covered all the other aspects above is to have fun with the exercise you are doing. Ensure that the activity that you have chosen doesn't make you feel bored and not fun. When you keep things energized you less likely to give up. So, for this, try to join a class or get a workout partner as a start to push you and keep you occupied.

CHAPTER 5

CLEAN OUT YOUR CLOSET

By this time, you've probably heard of Marie Kondo and her method of decluttering and practicing minimalism. And thanks to her and many others before her, plenty of individuals on the minimalist movement has pursued this way of life for several reasons.

HOW CAN DECLUTTER MAKE WAY FOR SELF-LOVE?

Plenty of people on the decluttering movement do this to consume less and to reduce their carbon footprint on earth, some to focus on health, some to discover their purpose and mission in life, some to pursue their passions, eliminate discontentment and some do it to contribute and create more to the society and community around them.

The concept of decluttering is to look or attain happiness not through material objects but through life itself. And that's all we want in life, isn't it? To be happy? To self-love yourself is to also remove material obsessions.

Happiness, just like the reasons for decluttering is very subjective and personal. Different people find happy in different things. To some, being close to friends and family is happiness whereas to some, working with children or animals is happiness no matter what the salary is like. To some people, being successful in their careers is what they like best. So really, it is up to us to find what is necessary to feel happy and what is just clutter. Clutter can be in the form of many things and more often than not, to feel self-love is to REMOVE the clutter, discard it, give it away or throw it. And this clutter can be in the form of items, objects, relationships that are toxic as well as habits that bring you down.

- Have a soft cashmere sweater given to you by an ex that you can't seem to let go?

- Can't seem to find what you are looking for in all that junk?

- Closet overflowing because there are just too many things?

- The relationship with your college friends is becoming depressing?

Take note that there is absolutely NOTHING WRONG with material possessions. However, in practicing self-love, it is also about moving toward living a more clutter-free life and like all women (or most women) plenty of our clutter is found in our closet. Our world today has become extremely consumerist and materialistic that we end up assigning our things and give too much meaning to them. In that process, we end up forsaking our health and relationships, our passions and personal growth. Decluttering your closet and living spaces isn't to say that you cannot own a car or a house or those pair of Manolo Blahnik's. It also says that you can't raise a family or have a career.

It simply means minimizing the fluff around you so you can make better decisions consciously and more deliberately.

DECLUTTERING TOWARDS MORE FREEDOM

Decluttering is getting rid of anything that is unused, unnecessary and unloved. It involves items, things, objects and sometimes even unhealthy obsessions and practices. Decluttering is all about shrinking the things you own and connect with by eliminating the excess. When you are done with decluttering, the items you own or the practices you continue to follow are the ones that you continuously use, love and are healthy.

GET RID OF THAT 'THIS STUFF COULD BE USEFUL ONE-DAY MENTALITY'

One of the reasons why we keep stuff is because we think we'd used it someday. Go on, think about the time your mother or father or grandparents or even a friend has said "I'm going to keep this. I'd need it someday', only for that day not to come or the stuff is gone that one time we need it again. Yes, these things that you do not like and neither use could be useful someday, but we often let this fact confuse us into keeping all sorts of things. The question that you should be asking is: is this thing necessary to me now?

Something that is useful does not mean that it is necessary. It's useful to have a blow torch but it is necessary.

Could you make do with something else that you commonly use? Would you be able to borrow this item from friends and family when you need it? What would happen if you wanted to use this thing and you didn't have it? If you are expecting an answer for all of these questions, you are not going to get one. The answer really depends on the item you want to remove. It also depends on where you live, your access to friends and family as well as maybe your occupation and what you do on a daily basis.

You may need a blowtorch to brulee your sugar or broil the top of a lasagna but how often do you make this? If you cook and bake a lot and you've used the blow torch a couple of times, keep it. But if you've only used it once, you can actually give it away to someone who would really use it. Instead, you can use the broiler of your oven or even a lighter to do this.

Keeping a set of candles in case of a power failure is very different to keeping a pair of tennis rackets in case you ever decide to take up tennis. We need to be practical about what you keep and what you give away. You can also think about your own situations and circumstances and use them to make proper decisions about what is useful to keep and what is necessary to keep.

You do not want to keep things 'just in case it may be useful someday'. This will turn you into a hoarder. Pass them on to people who will use them more often than you would.

It is helpful for us to try and understand what decluttering means otherwise we will end up creating plenty of reasons and excuses for why we do not get rid of the things that we do not use. We also end up purchasing more things then we need, and this makes our living spaces, our offices and our life so unorganized and cluttered.

CLEANING AND CLEARING CLUTTER

THE METHOD OF DECLUTTERING

Start by working from room to room to sort items into their appropriate boxes. Work with one item at a time, one room at a time. Go through your closets, storage spaces as well as your cabinets. Remove whatever clutter you find in your living spaces, home offices, bathrooms and kitchen. Be sure to throw out the necessary trash immediately. Box up the storage box when you are done and place the giveaway/sell box in your garage. If you keep looking at your boxes, you will end up tricking your brain to rescuing the clutter you have kept away.

Things to Remember

- Pruning items with your emotional value can be difficult.

- Consider them carefully if you need the item to have an emotional attachment

- If you are not sure what boxes these items go to, consider storing them on a trial basis or give it to another family member

- You can also take photos of these items if you feel like you need proof of memory

- If you feel strongly about keeping these items, then do not fight it but instead incorporate these items into some kind of display for your home or in organized storage.

- Get rid of clothing you cannot fit in in the hopes you would return to your size before. If you need motivation, keep one item but let go of the rest. You can reward yourself with a new wardrobe once you reach your goal weight

- Try not to keep sets of things. Sets of cutlery or glassware can be beautiful but space-consuming.

- Keeping things just because you need it someday will clog your closet. Remind yourself that you need the space.

DECIDE TO TAKE ACTION

As the saying goes, Rome wasn't built in a day and as such, it didn't take a day for our homes to fall apart and it definitely won't take a day to piece it all back together. And that is okay. The most important aspect to consider here is taking that first step in dealing with the homemaking issues and taking it one step at a time. Each step, no matter how small or big will be one step better in resolving the issues we have.

Amongst one of the best first steps that we can take is putting aside one hour each day to step away from our electronic appliances. And when I say electronic appliances, I mean your phone, computer, notepad, television as an example. Just like pulling a Band-Aid off a wound, this one hour rule will be challenging at first, but as time goes on and if you keep to this challenge you will find that you are able to accomplish so much more than what you would have done when you aren't scrolling through videos on YouTube.

The next thing to consider is never comparing your home against someone else's. Each home and families have varied challenges, situations, limitations and starting points. Do not let those near-perfect images in magazine articles make you feel that there isn't any point in what you are doing. Never fall into this trap. Your kids and husband will not feel any less loved or feel neglected if your home isn't as organized or pristine as the one-off an Ikea product catalog.

Sometimes, even with the best attitude and character that we have, the obstacles that lay before us can be daunting and mind-numbing. Playing catch up will be a lot harder than keeping up in some cases. The pathway to getting the household to a pristine environment will be challenging but you will find that maintaining that level of cleanliness from time to time will eventually get a whole lot easier.

Here are some pointers on where to start:

1. **Pay special attention to time-constrained tasks and issues**

- Relook for any task or activities that would have slipped your mind

- Identify bills that need to be paid

- Identify tasks and activities that are associated with time (i.e.: doing your laundry, sending in your car for service, etc.)

2 **Focus your efforts on food and clothes first**

- Make a menu of all the meals that you will prepare for the entire week not only for yourself but for your entire family.

- Prepare a meal plan for each day and decide which day will you go grocery shopping

- Get your laundry done at one go

- Focus on getting your meals prepared for everyone your laundry load for two days done.

3 **Plan your morning routine**

- Make a list of all things that you need to get done before lunchtime

- Tick off each action item once it's done.

- Make sure to plan and prioritize important actions.

- Don't plan items that are not essential to the smooth running of the day.

- Start will small activities and don't worry if you find one or two activities going past beyond the afternoon. You'll start seeing progress pretty soon.

4   **Plan a simple afternoon and evening routine**

- Make a list of all things that you need to get done to ensure a smooth-running day (i.e.: laundry, cleaning, lunch and dinner prep)

- Start small.

- Tick off each action item once it's done.

- Make sure to plan and prioritize important actions.

- Don't plan items that are not essential to the smooth running of the day.

- Start small and get comfortable with the basic tasks before adding on harder tasks as time goes on.

- Take things slowly. A few tasks in the morning, afternoon and evening should do the trick. Don't try to bite more than you can chew.

- Focus and don't get discouraged. Expect to see results over a few weeks.

- Never expect immediate changes and results. That's a recipe for disaster.

- Keep working at these tasks and work towards being efficient in completing them.

5   **Identify one special task to do each day**

- Pick one weekly task to accomplish each day. This can be from vacuuming to mopping the house.

- Set a time and work on a particular area that needs decluttering or cleaning.

- The objective here is not about perfection but accomplishing the job within the timeline set.

Keep this mantra or motto in mind when you got about these tasks, "Better today than we were yesterday; better tomorrow than we are today." Take small steps and keep working at decluttering and cleaning your home each day. As it goes, you'll start making progress and achieve your goals and objectives in the long run.

WHEN IN DOUBT, START WITH THE FLOOR

It is not always necessary to have a clean and spotless floor, but there are some reasons to why starting there helps. This is because it makes the most difference when you're done with it and it won't take much out of you

This is because the floor is the place that gets messy first and quickly compared to any other part of your house. It gets even messier when you stop tending to it for even a short period, especially when you have babies and kids moving around. Things get left on or moved on the floor and if not cleared or stored it becomes cluttered very quickly. Having a messy and cluttered floor tends to create a very stressful environment within a home. Thus, have a clean and clear floor is one sure-fire way of reducing the stress and one step to a decluttered and cleaner home.

You will notice right away that cleaning the floor is the least labor-intensive task as compared to the other tasks that you will undertake in your household. Once you start having one portion of the floor cleared and cleaned, you will start to have fewer amounts of stress and feel a little less overwhelmed. And hopefully, this will start to inspire you to keep going on further in this decluttering journey. Even if it doesn't get you all inspired and raring to go, you still managed to get some cleaning done and a little progress is so much better than no progress at all.

VALUING SPACE OVER STUFF

Yes, sure some people work better or function more productively with chaos, or should we say controlled chaos. A little bit of clutter helps some people think better or perform better. Too much organization and not enough clutter may seem a little bit unnatural or robotic. However, most people can agree that the most hated kind of clutter is pointless clutter, not controlled clutter. This pointless clutter is just things that keep piling up from laundry not folded to unanswered emails to boxes that you plan on donating to the shelter or even the empty pots you plan on planting your flower seeds on or even those beauty samples you got in gift bags but never used. It could even be clutter which you think is useful like those books you bought but never read or even those computer parts you purchased but never got around to fixing it up.

This is the kind of clutter that makes anyone crazy and stressed up. One minute you have one magazine on the table and you think 'I will get to it later' only to see in three weeks' time, there are another 4 more magazines on top of it left by someone else in the family, or even by you. The average person hates unnecessary and pointless clutter. But the issue here is that this pointless clutter slowly accumulates over time, you barely realize it exists until suddenly, you find it piled high up in your spare bedroom or the storeroom.

Each time you walk by it, you tell yourself that you need to get this sorted out but the thought of it makes you feel overwhelmed. When will you do this? Will you have the time? How long will it take? Where do you start? The thing is, decluttering doesn't have to overwhelm at all. When you look through your stuff and sort them out, you are probably thinking about what you are going to keep and what will you give away, what can be sold and what needs to be thrown. Most of us have one basic question: Is this item worth keeping?

As we talked about in the chapters before this, the keyword here is WORTH. What matters is what this object or process or system is worth to you and to the members of your household. Certain interests need to be taken into account for sure. Apart from figuring out an object's worth, you also need to look at the amount of space you have in deciding to keep or throw things away. So, there are a number of ways to come up with this answer. You can look at it as a mental trick or you can look at it as part of imagination. We'll consider five of these little mental tricks:

1. Imagine you don't own it and are buying it.

This mental trick is courtesy of Carleton University, suggesting you look at things on whether you will want to pay a certain price for it. If you had not already owned the item, would you pay a certain amount for it? If you answered NO to this question, sell it immediately. If you do this practice for all the items in your house, you will come to the realization of how little the value you place on some things that you first thought you couldn't part with. All this is done by imagining when you do not own it so that it gives you a better idea of its value.

For instance, you own a box of cigars that you plan on getting rid of. You go on to eBay and find out how much this type of cigars are sold for. It gives you a good idea of what you can get if you sold your own box. Then you ask yourself, would you be paying that much if you didn't already own it? You can also pretend that you lost a certain item in a house fire. Would you risk your life to save it right away or perhaps sometime later? If you do not get it urgently, then you know the value is low. But if you urgently wanted to get it, the value is obviously high.

2. Imagine you're moving, and hiring a professional mover

What's worse than packing? Having to pack and move. Moving homes is not only a pain, but is also a hefty price tag. So, when it comes down to the value of your things versus the value of your space, here's another question to ask yourself: how much would you pay to move this item across the country?

3. Calculate the cost to store the object

Here is some simple math that you need to do. First, you need to look at the cost of your living space and then calculate that with the proportion of the area that is occupied by the stuff you don't use. You can use the price you paid for the purchase of your home as a figure or if you know the current estimated value, better still.

So, if you paid $500,000 for your home and you used 20 percent of its area for impractical stuff. That means you are paying $100,000 for storage of space. This amount plus the cost of heating and cooling is definitely going to increase your bill so just imagine you actually rented space to keep this stuff.

When you think of things in this kind of perspective, you can definitely see a more objective approach towards keeping or getting rid of unnecessary clutter. Here is the following formula to calculate what each square foot of your house is worth:

Value of your home ÷ Square footage of your home = Value of each square foot

When you know the value that each square foothold, you know what to do each time you see a square of clutter. Put a price tag on it and see how it costs you.

4. Look into the emotional cost

Still not convinced that you need to declutter. That's okay. Now we weigh the emotional costs associated with clutter. Financial costs are not the only ones related to our things. There are also emotional costs connected to it. While you may not be able to set a dollar value on these items, it carries significant weight on your decisions. How do you feel when you hold a certain item? Do you feel sad, or happy? Does it bring back certain memories? What memories do you want to feel when you touch or carry an object? How you feel will tell you whether to keep or toss something.

5. Estimate the cost of the things

Another mind trick to use is to estimate the cost of things you purchase because the ones you already own are hidden somewhere in your clutter of things or you've forgotten about them. So, what do you do? You declutter your home so that you know what you have and where it goes. Apart from the five mind tricks listed above, here is another decluttering formula you can try, and it is in an acronym of RFASR.

- Recency—When was the last time you used this?
- Frequency—how often do you use this?
- Acquisition Cost— How expensive is it to get this again?
- Storage Cost—how much space cost does this item take?
- Retrieve Cost— what is the cost of retrieving this or has this item become outdated?

Once you have this in your mind, here is the equation for you:

R (Low) + F (Low) + AC (Low) + SC (High) + RC (High) = Not Worth It

When using this formula, you can follow the flow like this:

- Recency: I used this two years ago
- Frequency: I did not wear this a lot back then
- Acquisition Cost: I can always order this online. It costs less now than it did before.
- Storage Cost: This is going to take a lot of space in my garage.
- Retrieve Cost: It's so last century…

In situations such as this, you get rid of the kinds of things because it does not add value or usefulness. If there is an emotional attachment to the things you have, and you want to remember it, then you've achieved one goal which is to keep it. However, for some reason if this item takes up space you cannot afford to keep, then you may want to change the connection to the gift.

CHAPTER 6

SELF COMPASSION IN YOUR SELF-LOVE JOURNEY

You're so dumb! You don't belong here loser! Those jeans make you look like a fat cow! You can't sit with us! It's safe to say we've all heard some kind rude, unwanted comments either directly or indirectly aimed at us. Would you talk like this to a friend? Again, the answer is a big NO.

Believe it or not, it is a lot easier and natural for us to be kind and nice to people than to be mean and rude to them whether it is a stranger or someone we care about in our lives. When someone we care is hurt or is going through a rough time, we console them and say it is ok to fail. We support them when they feel bad about themselves and we comfort them to make them feel better or just to give a shoulder to cry on.

We are all good at being understanding and compassionate and kind to others. How often do we offer this same kindness and compassion to ourselves? Research on self-compassion shows that those who are compassionate are less likely to be anxious, depressed or stressed and more resilient, happy and optimistic. In other words, they have better mental health. In this chapter, we will explore self-compassion and how you can work towards becoming more compassionate to yourself and towards practicing in your self-love journey.

UNDERSTANDING SELF-COMPASSION AND ITS BENEFITS

What is self-compassion? Have you thought about it or experienced it from someone? The truth is, having compassion for yourself is not different from having compassion for other people or animals. Having self-compassion is being kind to yourself and understanding to your needs when you face personal failures. Think about how you would talk and console a friend who's going through a rough time: what would you say to them? Would you be harsh to them? Would you say things that bring them down even more?

The answers to those questions are of course a big NO. You would do what all good friends do, bring them up when they feel down, hug them and tell them everything is going to be ok, telling them that you'll be there for them to talk to or if they need help. Self-compassion is acting this way towards yourself when you go through a rough patch. You notice the suffering and you empathize with yourself by comforting yourself, offering kindness and understanding.

Kristin D. Neff and Katie A. Dahm are two prominent are two names synonymous with research on self-compassion. In their book, the Handbook of Mindfulness and Self-Regulation, it states that there are three primary components to self-compassion:

1. Self-kindness

2. Common humanity

3. Mindfulness

To understand self-compassion, we need to consider what it means to feel compassion on a general level. Here are some views of compassion:

- The Buddhist point of view of compassion is given to our own as well as to others suffering.

- Goetz, Keltner & Simon- Thomans, 2010: Compassion is the sensitivity to the suffering that is happening, coupled with a deep desire to alleviate that suffering

- Neff, 2003a: Self-compassion is compassion directed inwards, referring to ourselves as the object of concern and care when we are faced with an experience of suffering

THE THREE ELEMENTS OF SELF-COMPASSION

The key to understanding self-compassion is to understand the difference between this trait and more negative ones. Sometimes when we give ourselves self-compassion, it may be construed as narcissism to a point, which is why it is important to know what self-compassion is and to what degree is it considered self-compassion and when it isn't.

1. SELF-KINDNESS IS NOT SELF-JUDGEMENT

Self-compassion is being understanding and warm to ourselves when we fail, or when we suffer or at moments when we feel inadequate. We should not be ignoring these emotions or criticizing yourself. People who have self-compassion understand that being human comes with its own imperfections and failing is part of the human experience. It is inevitable that there will be no failure when we attempt something because failure is part of learning and progress. We will look into how failure is a friend in disguise in the next chapters. Having self-compassion is also being gentle with yourself when faced with painful experiences rather than getting angry at everything and anything that falls short of your goals and ideals. Things cannot be exactly the way it should be or supposed to be or how we dream it to be. There will be changes and when we accept this with kindness and sympathy and understanding, we experience greater emotional equanimity.

2. COMMON HUMANITY AND NOT ISOLATION

It is a common human emotion to feel frustrated especially when things do not go the way we envision them to be. When this happens, frustration is usually accompanied by irrational isolation, making us feel and think that we are the only person on earth going through this or making dumb mistakes like this. News flash, all humans suffer, all of us go through different kinds of suffering at varying degrees. Self-compassion involves recognizing that we all suffer and all of us have personal inadequacies. It does not happen to 'Me' or 'I' alone.

3. MINDFULNESS IS NOT OVER-IDENTIFICATION

Self-compassion needs us to be balanced with our approach so that our negative emotions are neither exaggerated nor suppressed. This balancing act comes out from the process of relating our personal experiences with that of the suffering of others. This puts the situation we are going through into a larger perspective.

We need to keep mindful awareness so that we can observe our own negative thoughts and emotions with clarity and openness. Having a mindful approach is non-judgmental and it is a state of mindful reception that enables us to observe our feelings and thoughts without denying them or suppressing them. There is no way that we can ignore our pain and feel compassion at the same time. By having mindfulness, we also prevent the over-identification of our thoughts and feelings.

BENEFITS OF SELF-COMPASSION

You've probably heard your parents say time and time again to treat others as you would want them to treat you. Therefore, we are often taught to be empathetic and compassionate to others who are facing difficulties and challenges in their life.

However, when faced with our own personnel challenges be it in our everyday lives, work and relationships, we often find ourselves becoming our own worst enemy. Hence, we become too critical and judgmental on our own selves and in turn prevent any healing process from taking place. Therefore, instead of being self-critical to oneself, we need to develop the concept of self-compassion in combating our negative thoughts and self-criticism that keeps us from overcoming our obstacles and challenges.

- Self-compassion is defined as being compassionate to our own suffering, inadequacies, weaknesses and failures. As we know from the previous chapter, Kristin Neff, an associate professor at the department of educational psychology at the University of Texas further breaks down self-compassion to 3 key elements which are self-kindness, common humanity and mindfulness.

- Self-kindness is about recognizing our flaws and issues as well as being caring to oneself when going through bouts of hardship and challenges. Common humanity, on the other hand, puts emphasis that the suffering and anguish we go through are all a natural part of being human and it's a normal part of everyday life. Lastly, mindfulness deals with the individual's ability to take a middle path in addressing their sufferings so as not to neglect or overthinking the situation.

Various research done on the topic of self-compassion indicates that individuals who practice self-compassion have far greater psychological health than those who lack it. The individuals who practice self-compassion have more positive life satisfaction, happiness and optimism. Apart from that self-compassion is also connected low levels of anxiety, self-criticism and depression. As such, in a way, self-compassion can be used as a tool to develop inner strength when facing challenges in every aspect of our life.

So, we know what self-compassion is and sure it helps us lead a better life and have better relationships. What other aspects of self-compassion are there?

Here are some major benefits you can reap from being self-compassion. We explore it in terms of work, relationships and life.

SELF-COMPASSION AT WORK

Our daily work environment can be a long-lasting love-hate relationship with its own ups and downs that one has to face on a daily basis. As such, we are constantly bombarded with undue stress in meeting deadlines, reports and customer expectations. Many at times, we will face moments that completely overwhelm us and have a negative impact on us. This can be caused by numerous factors such as a negative remark by a colleague, superior or even a customer, failure to reach sales targets or goals, not getting that raise or promotion that you so deserve or even by making an unintentional mistake at the job. Since we all strive to achieve more and be perfect at our jobs, these negative circumstances will have an adverse effect if not dealt properly and swiftly.

Self-compassion can be used at work through the following means to reap various benefits:

- Conducting a post-mortem to review the shortcomings and failures of a certain project or task and learning from these failures to prevent similar occurrences in the future.

- When facing criticism and rejection from colleagues, superiors and customers, instead of being self-critical and falling into complete despair, we will be able to be calm and focus our energies and thoughts of improving ourselves and not to allow stress to overwhelm ourselves.

- Applying self-compassion at work also helps us in being resilient through difficult scenarios especially is situations that we don't get a certain reward or promotion that we think we deserve.

- Self-compassion enables us to be more creative. When we fail a project or we do not complete a task or when a work event doesn't go as expected, being self-compassionate to ourselves will help us to look back at the series of events and instead of berating ourselves, we look back and see what we could have done better and learned from our mistakes. It makes us become more creative the next time around.

- Self-compassion builds trust. It enables you to be transparent and authentic, makes it easier for people to connect with you because you are your true self.

- Showing genuine compassion to yourself also means showing compassion to the people around you. When you show compassion to yourself, you extend this feeling to your co-workers and it makes them feel safe.

- Self-compassion allows you to allow yourself and your team implicit permission to do their very best without the worry of punishment or repercussions if something doesn't go right.

SELF-COMPASSION IN RELATIONSHIPS

In the topic of relationships be it a romantic or non-romantic relationship, we often find ourselves in situations of disagreement from time to time. And these can sometimes lead to moments of stress and unhappiness between oneself and their significant other/parent/sibling/friend. Self-compassion provides various ways much like our situations at work to help us deal with these issues and challenges. Many studies done on this matter point that self-compassion, when used, have the following positive impact on relationships:

- Individuals who practice self-compassion know that every individual as well as themselves aren't perfect and are subjected to weaknesses and shortcomings

- They are able to relate to their partners much better

- They are warmer and more compassionate in understanding a situation

- They are more open to compromise to resolve a situation

- Individuals who are self-compassionate have better empathy. They bring out the best in their partners.

- They are more responsive and aware of the issues that their partner faces

- They are better listeners, they listen to understand and not answer

- People who practice self-compassion own up to their mistakes

Studies also have shown that individuals that lack self-compassion tend to have a negative effect on people around them which may lead to isolation. As such, those people who practice self-compassion have healthier and happier relationships and have a bigger wide social circle.

SELF-COMPASSION IN LIFE

When encountering difficulties in daily life which can range from a number of issues/aspects such as health to financial issues, we need to act by being compassionate and kind to ourselves. When faced with various issues on a daily basis, self-compassion allows us to look for solutions to take care of oneself instead of berating or being overly critical of one's lack of accomplishments or weaknesses. With that being said, an individual who practices self-compassion will look into various ways to engage their mind and body into healthy activities that can stimulate them and lets them focus on the positive aspects instead of groveling on a negative situation. This can be in the form of an exercise, a hobby, prayer or even a warm bath or a cup of tea to calm themselves down.

Self-compassionate individuals tend to be more:

- Happier

- Satisfied with life

- Resilient

- Emotionally intelligent
- Have better-coping mechanisms
- Optimistic
- Creative
- Less judgmental
- Better goal-getters
- It greatly reduces mental problems

As such, cultivating the habits of self-compassion in every aspect of our life will allow to become the best version of ourselves and allow us to live much happier with the right mindset.

Dealing With Negativity

Did you ever realize that it is much easier to be happy than it is to be unhappy? Go ahead. Think about it. While you are reading this, just think about the many things that happened before you open this book and reading. What happened when you woke up? Did you get a kiss from your partner? How did your coffee taste this morning? How is the weather outside like now? All these things that happened to you today, what made you happy and what made you sad?

If you listed ten things today and 7 of them were things that made your happy and three made you unhappy, sad, frustrated or moody, then most likely you were grateful, and you were positive. The thing is, many of us would prefer to be happy and positive rather than be unhappy and negative. And it is that simple to be positive and happy.

Also, positive thinking is above and beyond just being happy or displaying a cheerful and upbeat attitude. It also creates and establishes value in your life and relationships, and it also helps you build skills that benefit you longer than your smile can take you. Barbara Fredrickson, a positive psychology researcher from the University of North Carolina, published a landmark paper on the impact of positive thinking on work, health and general wellbeing.

Here's a little brief of Barbara's research:

What Can Negative Thinking do to Your Brain?

Our brain is programmed to respond to negative emotions by shutting off the world around us and limiting the options we see around us. For example, if you get into a fight with your sister, your emotions and anger might consume to the point where you react adversely, you can't think about anything else. Or for instance your coffee this morning spilled on your shirt, and this creates a domino effect of everything going wrong in your day, and you get so stressed out that you find it hard to start or do anything because you've lost your focus. Or if you are supposed to complete a project but you didn't, you start to feel bad about it and all you think is how irresponsible you are and that you are lazy, and you lack motivation. The point is, our brain shuts off from the outside world and relies on the negative emotions of fear, stress, and anger. Negative thoughts and emotions prevent us from seeing other options, solutions or choices that are around us.

What Can Positive Thinking do to Your Brain?

Barbara Fredrickson also explains how positive thinking manifests in our brains. She explains with an experiment where research subjects are divided into five groups, and each group is shown a different video clip. The first group was shown clips that created feelings of joy, whereas the second group was shown clips that created contentment, the third was the control group that had images of no significant emotions and were neutral whereas group four had clips that created fear and group five had clips that created the feelings of anger.

Participants were then asked to imagine themselves in situations that these same emotions would come about and write down their reactions to it. Participants viewed images of fear and anger had the least responses or reactions whereas participants who saw joy and contentment had more reactions. The bottom line is, if you experience positive emotions you will see more possibilities in life. Positive emotions broaden our possibilities and thinking, thus opening up more options for us in facing issues, crises, problems, and solutions and so on. In the next few chapters, we will discuss how we can work our minds to be more positive and look at things in a more positive perspective to enhance and give more value to our life, relationships, and goals. It is not as hard as it seems because all it takes is a little practice.

Have you seen the movie Inside Out?

If you did, then you will probably realize that being sad is a good thing, not always, but this emotion is there for a reason. When we talk about dealing with negativity, it doesn't necessarily mean being optimistic all the time, especially in the face of suffering.

Pain and sadness are just part of the complex human emotions all of us have, and it is just as important to feel pain and sadness, guilt and fear as this are all part and parcel of coping. Experiencing and processing negative emotions in a healthy way is a crucial part of personal growth.

There are two scenarios when people are confronted with negative situations. One, they either obsess over the problem or two, they numb their emotions. Either of these coping methods are not healthy, and it can create harmful patterns in our mind, over a period of time. Obsessing is deceptive because it feels as if you are thinking things through but to continuously obsess over a situation only reinforces the impact of the negative thoughts and emotions. That said, numbing your emotions towards a pained situation isn't good either because it really is not possible to selectively numb out an emotion. Humans are so complex that our range of emotions does not enable us to directly shut down an emotion. If you somehow blot out anger, you'll blur out happiness and serenity too. Why? Because while you like being active and optimistic all the time, not showing anger to something that has hurt you or hurt you or frustrated you, will make you feel more bitter eventually. Only because you weren't able to express your anger, the situation or the person related to this will not know how you feel. For example, if we use alcohol to numb our pain, we do not learn how to cope with sadness. We just develop another problem which is alcohol abuse.

If you are going through a painful time, then you need to develop healthy coping skills, and this involves recognizing the inevitability and necessity of some suffering and moving on from it. The process usually includes:

- Acknowledging your negative feelings and watch them with a non-judgmental attitude

- Recognize when they are triggered and assess your reactions when responding to this

- Understand that pain is just a catalyst for growth and resilience

- Practice forgiveness towards those who have pained you

- Express yourself in creative and healthy ways like painting or exercising

- Seek the support of others

STEPS TO DEAL WITH NEGATIVE THOUGHTS AND EVENTS

Here are some tried and tested ways to overcome negative thoughts and events which you can try:

1. Meditate or do yoga.

Yoga helps take your focus away from your thoughts and bring attention to your breath. Yoga or meditation is very relaxing, and it helps ease one's mind. It also helps you stay present and focused on the moment that is happening.

2. Smile.

Pain and sadness can make it very hard to smile. While it does seem hard to smile when you aren't feeling so happy inside, you need to sometimes force this out of you. So, try doing this in front of a mirror every day or make a mental note to smile to the people you correspond with daily.

3. Surround yourself with positive people.

Surround yourself with friends and family that can give you constructive and loving feedback. Each time you feel you are going down into your negative spiral, call these people up and speak to them so they can put your focus back again to where it's needed to be at the time.

4. Change your thoughts from negative to positive.

Easier said than done, no doubt but you can turn any situation into a positive one. For example, if you have just started a new job you barely have enough experience of, instead of saying 'I'll take a long time to adjust or learn' just say 'I will take on any challenges because challenges excite me!'

5. Don't wallow in self-pity. Take charge of your life

You are the captain of your shop so do not make yourself a victim. There is always a way out of any situation so if it becomes unbearable, then leave. Otherwise, you stay put and make the best of it and don't point fingers, blame, complain or whine.

6. Volunteer

Volunteering also takes the focus away. If you think you are in a bad situation, imagine the people who need food aid or money. Do something nice for someone else so volunteer at an organization or donate.

7. Remember to keep moving forward

We easily dwell on our mistakes and feel terrible for the way we acted. But you can't reverse the situation so instead of feeling sorry for yourself or beating yourself up over what you'd done, tell yourself that you'd made a mistake, you learned from it, and you want to move on.

8. Listen to music

One of the best ways to alleviate your mood especially in the morning is by listening to songs and singing in the shower! It doesn't matter if you remember the lyrics – a good happy song will put you in a good and happy mood.

9. Be grateful

Being grateful enables you to appreciate all the things you have. So be grateful every day.

10. Read positive quotes.

Just log into Pinterest, read positive quotes every day. Better yet, print out the ones that you like and stick in on your wall, your fridge or your computer.

LEARN TO FORGIVE YOURSELF

Just like negative emotions, failure is also good to experience because it only makes us stronger.

Yes! Failure is something that didn't kill you. You're still alive! So, what doesn't kill you only makes you stronger. Why do you need to experience failure? Nobody wants to experience failure but if you looked at the successful people in our generation today, or even the past, they all failed. They all made mistakes. They all went through trial and error. What sets them apart from the perennial failures? They didn't give up. They learned from their mistakes. They had extreme passion, making them eager to keep on trying until they succeeded. Here's why you need to experience failure:

- Without failure, you'd be sucked into a blissful feeling that nothing can go wrong and that everything you'd put into place will work exactly as how it should be

- When something does go wrong, you are unable to cope with the change or adapt to create solutions.

- Failure enables us to work on our flaws and it also allows us to right our wrongs. Failure also enables us to upgrade or enhance or refine our work, technique and solutions

- Failure also teaches us a lesson. It is our choice to learn from it or run from it

When we fail, it's easy to get discouraged and upset and we develop a sense of being afraid to fail again. In order to be successful in anything that we do, we just need to remind ourselves to let go of our pride and ego. Failure only makes us grow wiser, make us more adaptable and vivid to any possible scenarios that could happen. We are more prepared to face the same problem but at a different angle.

So how do you look at failure in a positive way? We need to redefine the way we view failure. The fear of failure is what stops many great individuals from creating something beneficial and meaningful in our world. The fear of failure is why we stop ourselves from living extraordinary lives. The fear of failure is why we never submitted the novel we wrote; we never expressed our feelings to the people we love, never bungee jumped or telling someone how you really feel.

Daniel Epstein, founder of Unreasonable group stresses the point of re-branding the way we see failure. He suggests defining it as such:

"To Fail means "to not start doing something you believe in. To stop doing something you believe in just because it is hard. To ignore your gut instinct around what you believe is right and wrong."

In actual fact, many of the world's greatest philosophers, entrepreneurs, scientists and artisans have all expressed their thoughts on failure and how it has helped them overcome adversity and obstacles. All these perceptions tell us that fear of failure is evident in every human being but with passion and perseverance to achieve what you want will be the driving force in the determination between constant failure and success.

Chapter 7

EXPLORE YOUR SPIRITUALITY

As we know now, self-love like anything else needs to be actively practiced. For some of us, self-love is not given which is why we should always practice through our self-love rituals, habits and also through discovering or unlocking our spirituality.

Some people relate spirituality with religion while some do not. What you choose to align yourself with is entirely up to you. But in this chapter, we will focus on engaging with our spiritual side through meditation. When we do meditation (and even yoga), the primary factor to concentrate on paying attention to the NOW. We stay focused on what is actually happening rather than try to fix something or criticize something or even analyze something. We practice, through meditation, catching the NOW with an open hand.

In a lifelong practice of meditation, we learn to observe what occurs in our minds and how do we talk to ourselves in those moments. Sometimes, it can feel like we have very little control over our thoughts coming in and out of our minds. The internal conversation with our little voice is that we are not doing it right, or we are not good enough. It is hard to choose our thoughts. However, we can choose our actions. The more we get better at being mindful, we more we are able to show what we feel, what we need as well as the different options we have.

When starting your practice of meditation in self-love, tell yourself:

'*I may not always feel like I deserve love, but I can behave as I do*'

It is challenging to change our thoughts and emotions but it something necessary to do in the path towards self-love. When our mean thoughts get the best of us, we can decide to do something about it, such as immediately getting up from wherever you are and moving somewhere you can get some air. Or you can just walk away to get something to drink like water to hydrate yourself in order to get better clarity. Training our actions are a little harder compared to training our minds.

When we start to believe and behave to love ourselves unconditionally, we:

- Send messages to our subconscious minds to actually love ourselves

- We share signals to our mind that we deserve the kindness we experience

- We are more self-compassionate to ourselves

- Recognize when our voices are getting cruel

- Start to see when we are being hard on ourselves unnecessarily and when to actually push ourselves

Through meditation, we will be able to catch ourselves before we go down into an emotional spiral. We will also learn how to love ourselves better and realize that we do in fact love ourselves.

Self-love, for most of us, isn't innate. It's a practice. We can start right now.

Learning To Meditate

It's not just about reducing stress or letting go and taking a few moments for ourselves to try and recompose and regain our thoughts. Meditation is about more than that. It is about finding balance, inner peace and calm in a world where it seems almost every aspect of our lives is capable of triggering stress, worry or anxiety. Our bodies and minds may be strong and tough, but there is only so much negativity that it can take before it starts to take its toll and affect our health, sometimes to a point where it could become unbearable.

If only there were a magic formula of some sort where we could keep out these negative feelings that are capable of causing such destruction within our minds and bodies, but there isn't. Which is why we need to turn to meditation as a way of managing our worries and anxieties, to find a way to find that balance within ourselves and recharge our energy.

The beauty of meditation is that it is simple yet powerful. Simple enough that anyone can learn how to do it effectively with the right tools, teachings and techniques. Anyone can learn the art of meditation, and it isn't as difficult as you may imagine. Sure, you may have tried it a few times and found yourself struggling in the early stages to quiet your mind and achieve a focus, calm, and mindful state, but that is perfectly normal, especially if you're a beginner just starting out on this journey.

Mastering the art of meditation, like everything else, takes patience, time and practice. You're putting far too much pressure on yourself if you expect to get it right from the moment you sit cross-legged on your mat and shut your eyes hoping to achieve deep meditation right from the get-go. No, it takes time and practice, and you need to be patient with yourself. In this book, you will find a four-week plan that will help you achieve deep meditation, and the key to succeeding in this is to remember that you need to be patient. Practice makes perfect, which is why your goal to achieving deep meditation is spread out over four weeks, you need time to master each stage and phase of the process before moving onto the next. With repeated effort and your goal clearly in mind, you will see results at the end of the four weeks.

I'M READY TO START MEDITATING, HOW DO I BEGIN?
There's more to meditation than merely sitting cross-legged on your mat with your eyes closed as you breathe in and out, as you will discover when you progress throughout the rest of this book. There are many, many ways to meditate, but every meditation practice must begin with these most basic steps:

- **Step #1 – Know Your Intention.** You must first set your intention before you begin your meditation. You need to have a purpose in mind, to remind you of why you are doing this. This intention is something you need to think about before you begin each of your meditation session. Some examples of intentions could be that you hope to be less stress, you hope you gain better mental clarity, you hope to learn how to balance some intense emotions that are going on in your mind right now, or you simply want to take this time for yourself to quiet your mind after a long and stressful day (if you're meditating at the end of the day). Meditation is all about being mindful, so bring to mind before each session what your intention may be.

- **Step #2 – Posture Matters.** In meditation, your posture is important. You need to be able to breathe deeply and bring your attention inwards. You're going to be doing a lot of deep breathing during these meditation sessions, and you need to ensure your posture is right or you're going to find yourself struggling to get through a lot of pain and discomfort you might experience during the early stages of practice, especially.

- **Step #3 - You Must Relax.** The reason meditation focuses on deep breathing is because you need to let your body relax. Mindfully focus on relaxing every muscle of your body during your meditation session, from your face, neck and hands all the way down to your stomach area and other parts of your body. Being able to relax is key during a meditation session because you're not going to be able to learn to quiet your mind if your body is tense ball of anxiety throughout. Whenever you may find yourself tensing up, you need to remind yourself to relax, relax and relax.

- **Step #4 - Find Your Focus.** If you're struggling to steady your thoughts during your meditation, it helps to have something for your mind to focus on. The easiest thing would be to focus on your breath, to concentrate on breathing in and out slowly, rhythmically and methodically. Just focus on the in and out of your breath. During this time, you may find your mind wandering occasionally to other thoughts, and that's okay. Whenever you catch yourself being distracted, just slowly bring your thoughts back to your breathing once more. You'll get the hang of it soon enough.

ESTABLISHING A PRACTICE

As part of the four-week plan program to achieve deep meditation, you will need to begin by first establishing a practice for yourself. Meditation may seem like an easy enough exercise, but it is an exercise that is going to require you to be mindful of everything that you're doing through the session. Meditation has a purpose and a goal, teaching you to act with consciousness and to be mindful of everything that you do.

WHY ESTABLISH A MEDITATION PRACTICE?

Just like a home needs a solid foundation on which the house can stand upon, so too does your meditation sessions. That foundation is developing a meditation practice of your own. Without a firm foundation to stand on, it won't be long before whatever you're doing eventually crumbles and falls because there's nothing supporting it. That's just one way of describing how important it is to develop a sound meditation practice right from the very beginning of the process.

The purpose of establishing a meditation practice is because you want to make meditation a habit, a part of your daily life, something you are willing to do every day without even thinking twice or resisting it because you're pressed for time. Establishing a practice will help meditation become an ingrained activity in your life, much like how brushing your teeth or showering, preparing something to eat and even your daily commute to work. Those habits are so deeply ingrained in you that you do them without any effort or a lot of thought put into it. That's what establishing a meditation practice aims to do for you right now, and something you need to establish as a foundation leading up to your four-week plan program to achieve deep meditation. The following steps will help you start establishing a meditation practice for yourself, preparing you to be well and truly ready for the four-week plan to achieve a deeper state of meditation:

- **Baby Steps at First** – Trying to do too much too soon is how a lot of people crash and burn. You need to start small and this can't be emphasized enough. Yes, it seems like slow progress in the beginning, but that's okay. Remember that old saying slow and steady wins the race? Keep the bigger picture in mind and practice patience. Start small at first by meditating for short periods of time, maybe 5-10 minutes a day, especially if you're new at it. You can do anything for 5-10 minutes a day with no resistance, and the time will pass before you even know it. When you see how easy that was, it keeps you motivated to keep adding onto that. Creating small, achievable goals you can do every day is how you begin building the habit of making meditation a part of your daily life.

- **Make Use of Apps** – Smartphones are another thing that has become so ingrained in our lives, many of us wouldn't know how to survive without your smartphones every day. There is an app for just about everything these days, even meditation, so why not make the most of the tools you have to help you establish a successful daily practice? There are several apps available that could help you enhance your meditation sessions, with everything from timers to ambient sounds to help set the mood. If it helps make your daily practice more enjoyable, why not? You're more likely to stick to something if you like what you're doing.

- **Use Guided Meditations** – There are apps that will guide you through the meditation process, and if you need to rely on one, in the beginning, go ahead and do it. Guided meditations can be a great tool, especially for beginners on this journey, to help you stay on track and on the right path. It will help you make sure that you're breathing techniques are what they should be, that you're relaxed, it helps you visualize and it helps you free your mind and immerse yourself in the whole experience if you're not constantly focused on whether or not you're doing it right. Guided meditations make it much easier for beginners especially to start getting into the flow of things and helps you progress in the right direction with your meditation sessions, especially when you're doing it alone as a solo practice, it's good to know that you're heading in the right direction.

- **Create Your Space** – Creating a space within your home that you actually look forward to spending some time in is an essential part of establishing a consistent meditation practice. Create a space in your home that is dedicated solely for your meditation sessions and fill that space with anything you need to make you feel comfortable, which makes you feel like you want to be there for a while. You can fill it with pillows, cushions, pictures that inspire you, incense or scented candles if it helps, anything that helps soothe your soul and brings you a sense of calm. That will go a long way towards helping you make meditation a consistency in your life if you have a space that you look forward to spending some time in each day because of the comfort and calm that it envelopes you in.

- **Calendar It In** – In the beginning, you'll need to have reminders that you need to practice your meditation for today. Make it a point to pencil it into your calendar or make it note of it on your calendar app on your phone. It can be easy for other things going on during the day to take precedence over your meditation session, which is why you need to purposely make that time to just stop and meditate before the day comes to an end and you realize you didn't get to spend any time meditating at all.

BENEFITS OF SELF-LOVE MEDITATION

In this entire world, there is no one like you-you know that. There is no one who has the same exact thoughts as you or the same feelings, otherwise, this world would be a boring place to live in. You are unique and this at most times feels amazing but there will be times where you feel like being unique is the worst thing in the world.

Questioning "what's wrong with you" is a common response

It is tempting to dwell on these thoughts. The longer we dwell on them, the more things we will find that we are lacking in. It is also common to lack self-confidence sometimes and sometimes it is because of the world we live in, where it portrays confidence on the outside while on the inside, it is all a facade. This lie detracts us from our feelings of authenticity, and it makes us feel undeserving and unworthy.

But this is where self-love comes in. When we find our self in this mindset, we use approaches and rituals to remind ourselves why we should love ourselves and why this responsibility is on us and not on someone else. Self-love meditation can help us frame our thoughts towards becoming more grateful and appreciative of the fact that we are unique. This is something that self-love as well as self-compassion champions the ability to love yourself through all the flaws but also the ability to push yourself when you feel demotivated.

S<small>ELF</small>-L<small>OVE</small> M<small>EDITATION</small>

With self-love meditation, you meditate on feelings of both love and compassion towards yourself as a breathing, living miracle of yourself. Through the practice of self-love meditation (which again can be part of your self-love ritual), you will be able to cultivate the confidence to allow your individuality to shine, you will also be able to achieve your goals and to develop the courage to go out and enjoy life.

BENEFITS OF SELF-LOVE MEDITATION

Meditation, done regularly has been shown to:

- Reduce the tendencies towards self-criticism

- To lead a positive growth mindset in our confidence

- To enable us to view the moments that we feel inadequate

- Helps us to reduce anxiety and depressive thoughts

- Increase the amount of compassion for ourselves and for others

- Help increase our feelings of connectedness

- Lessen our feelings of sadness and isolation

- Experience more positive emotions

- Become more sociable

Your Past doesn't have to be your Future
The great thing about self-love that is enhanced by meditation is that it can rewire the brain and enable the brain to respond more positively to the positive habits and rituals that you begin to introduce and practice in your life.

According to research, you would be able to start seeing changes after only two weeks. Of course, provided you REGULARLY practice the rituals, affirmations and meditations.

You want to be happier, don't you? Being happy also means being confident of who you are and feeling less stressed. It takes work from you, and discipline to put in that work even for 5 minutes a day.

If life seems challenging for you, take heart in the knowledge that the work you are putting in self-love will enable you to take into your core and you will know that you do not need anyone's validation or opinions to feel beautiful and authentic. You can treat yourself to some self-love anytime, anywhere.

Even with meditating, you do not need a specific place to do it. 5 minutes in your car before heading into your home or office, or even in the shower is all it takes to embrace yourself and see your life with a set of new eyes.

This year is yours to give yourself the most wonderful gift of all to express how incredibly worthy you are, the gift of self-compassion and self-love.

CHAPTER 8

FOCUS ON SOMETHING YOU ARE GOOD

Self-esteem is all about how you feel about yourself that is determined by your feelings and also the evaluation based on your perception of yourself. A person with a healthy sense of self-esteem has a higher chance of success is anything that they set out to do. If you are one of those few who have strong self-esteem. Good for you!

However, it may be hard to understand those who struggle with it. Whether or not you have high or low self-esteem, keep reading this chapter because it's always good to reflect and it is always good to refine and restructure our thoughts with regards to self-esteem.

The person who is standing in the way of success more often than not is ourselves. When we do not believe in ourselves, we pay the price of our accomplishments and achievements, but all of these things are possible only if we believe in ourselves. Believing in yourself is that superpower you have within you. You need to realize that you are stronger than you seem, braver than you believe and smarter than you think.

This does not mean that you are expected to be positive and upbeat and have all the confidence and self-esteem turned on 24 hours a day, 365 days a year. It just means that you need to believe and trust yourself when you most need it, like walking into an interview or attempting something new you've never done before.

If you have low self-esteem all the time, it is like driving through life with a hand brake on. You don't try new things, you don't meet new people, you don't explore. To establish true self-esteem, we must concentrate on our successes and learn from our failures and push negativity aside. Strong self-esteem determines how we handle life to make sure your journey towards self-love and self-esteem is positive, self-assured and confident.

BOOSTING YOUR SELF-ESTEEM BY FOCUSING ON WHAT YOU ARE GOOD AT

LIST OUT YOUR ACCOMPLISHMENTS

If you have never thought about it, this is a good time. In order to find out what you are good at, you need to do a little soul searching and a good way to do this is to write down your accomplishments anywhere you would be able to see it everywhere. If you journal, then list down your accomplishments. Have a blog? Write a post about your accomplishments: just make sure it is in a place that you can see often.

Think back to all the accomplishments you have had throughout your life, think really deep and write down any small accomplishments you have from the exam results you achieved, the cross country you completed, the first meal you cooked all by yourself, the paper you got an A on, your first article in a newspaper, anything that you can think of.

These may all seem simple but as your list keeps growing the more you think about your accomplishments, the more you will start to see that not only have you accomplished quite a lot of impressive things, you are also capable of achieving more.

You might also begin to see patterns of the things you have done and when you felt really good about the things you achieved which in turn makes you feel good about yourself. Doing this exercise can help you tune into the parts of yourself that you never knew about and you will see that you are in fact capable and extraordinary!

SHARE YOUR EXPERIENCE

In the process of listing out your accomplishments, you will come to find that there is a pattern to what you like and what you are good at. See yourself joining and participating in archeology? Or you have joined several art lectures, classes and workshops and you have a pretty good understanding of painting? There will definitely be things that you know that other people do not and one way to boost your self-esteem is to share this experience or teach someone what you know. This simple knowledge that you may take for granted is valued by someone else seeking to learn. When you teach someone something you know, you not only share your knowledge but also prove to yourself that you are unique, and you also have the ability to share things with the world.

USE HEALTHIER MOTIVATION HABITS

At times, we will definitely feel like we need a pick me up and to motivate yourself to take action. Practicing regular motivational habits will help you stay on track with your self-esteem.

- Remind yourself of the benefits- a simple way to keep you motivated and have self-esteem is to remember why you are doing this and putting in all these efforts. If you need a reminder-write down the benefits of your journey and what you want to achieve from it.

- Keep doing what you really, really like to do, in the previous chapter, we talked about creating a 'What's working for me' list. Do that thing as often as you can so that life becomes a little easier to cope especially when the going gets tough.

Over time it will become a habit and your inner critic will pop up a lot less often.

TAKE A 2-MINUTE SELF-APPRECIATION BREAK
Stop, Look and Listen! We all need a 2-minute break, and this is one of those things that need to be practiced, giving yourself a 2-minute break every day and to try it for a month and see how it benefits you.

It just requires you to take a deep breath, slow down, close your computer and ask yourself a question: What are the three things that you most appreciate about yourself?

The things that you list out does not have to be big things. It can be little things that you realized today or yesterday. Realizing the things, you appreciate also makes you more grateful and you also increase your self-esteem a little bit more when you look back and reflect. It could be that you are glad you smiled at the security guard at your building or it could even be the 5 minutes you spent on listening to a coworker. Any small appreciative thing is valued and should be reflected. These short breaks do not only build self-esteem in the long run but can also turn a negative mood around and reload you with a lot of positive energy again.

EXPLORING YOUR PASSIONS
You've probably heard about the phrase 'choose a job you love, and you'll never have to work a day in your life'. While it is great advice, it is not that simple to follow. Many of us want to find a job

we love but as we all know, it is hard to find something we love easily. Also, we may already have a passion for a hobby or an interest, but it does not necessarily translate into a viable business for us or we do not have any intention to change it into a job and that's ok. If you want to explore your passions, here are some simple tasks and questions you can ask yourself to discover your passions or rekindle them again.

FINDING YOURSELF- IS THERE SOMETHING YOU ALREADY LOVE DOING?
By the time you get here, you probably have done quite a lot of lists related to what works for you and what you love doing. In a way, you would have already asked yourself about your passions.

Your passions can be anything and you may not have been aware of it until you ask yourself questions, dig deep and explore things. You may like crafting ceramics, painting, collecting quartz stones and even playing the guitar. Did you ever consider it as a possibility of turning your passion into a source of income? Many people have done this so there's no saying that you shouldn't. If there is something that you are passionate about, it's time to look really deep into it and see if it's something you want to make money out of. It is not necessary because some people have passions that they do not want to make money out of and that's fine.

REMEMBER WHAT YOU LOVED AS A CHILD
More often than not, our truest passions are the ones we did as a child but as life goes on and adulthood comes in, things change, and we forget our passions. So, this time, try to look back in your childhood and remember the things you did long before you have to worry about rent, about having food on the table and about your career. What were the things you did? Some of us loved science experiments or we loved swimming at the beach or even

fishing trips with our dad or Sunday baking with our moms. Get in touch with your inner child and try to remember the things you loved to do. It could be something you still love doing and this is another great exercise to get in touch with your passions.

ELIMINATE MONEY FROM THE EQUATION

We talked earlier on in this chapter about turning your passions into something that can make you money because then it won't feel like a job. Whether you chose to do this is entirely up to you. Some people choose to turn their passions into a money-making initiative whereas, for some people, their 9 to 5 job is a passion and so is rock climbing every weekend. Nobody is confined to have only one, true passion and not everyone wants to make money out of their passions. Some people just like doing their passion even without money because it's just a self-love and self-care ritual for them.

That said, the idea of eliminating money from this equation is for you to think of your passions in a different way. Most of us do not indulge in our passions too much because we can't afford to spend time away from work or to spend money on expanding our passions. If money was not an issue, would you be traveling right now? If money was not an issue, would you be busking with your guitar right now? If money was not an issue, would you be teaching English in the most remote regions in the world right now?

Money is always going to be a concern and the idea of looking at our passions without the monetary deal is so that you do not allow financial pressures to dictate your choices. Your career should lead to financial stability, but you should not allow financial security to be the defining factor in choosing not to pursue your passions even in the most minimal ways.

Love playing the guitar but can't go busking? Then join a church choir or a weekend group. Love teaching English to less fortunate children while being able to travel? Then sign up as a language teacher that has a programmed to teach in foreign countries.

If you will it, it will happen.

GET FEEDBACK FROM FRIENDS

If you looked into your childhood but still couldn't get an answer or if you eliminated the money issue and still don't have an answer, it could be that you are not the best judge of what makes you happy, it happens. Ask your friends and family members or just people who know you intimately enough to help you put a finger on your passions. Ask them when you are the happiest and what you do most enthusiastically. Their answers might reveal something you never saw in yourself and it could be a real eye-opener for you.

READ THROUGH UNIVERSITY OR COMMUNITY COLLEGE CATALOGS

You can also discover passions or interest by browsing through college catalogs that usually have summer courses which are a short term that you can take. What would you study if you could do it all over again? What were the subjects you were afraid to do before? Revisiting the possibility of becoming a student again can also point you in the direction of your interests and passions.

IDENTIFY YOUR HEROES

Whose career would you most want to emulate? Think of someone you know personally or professionally or in your extended frame of reference (from your bartender to Tyra Banks). Learn about how these people got to where they are right now or read about these interesting people, their career and life choices. Identifying your heroes can help you pinpoint what you want out of life too.

BOTTOM LINE

Once you are done with these exercises, think about where you are right now versus how you were before you begin. What have you learned? Doing these exercises takes a little discipline and commitment on your part. Take time off to do this and focus on what are the things you really enjoy, from spending time at the dog shelter to painting sceneries, to running a 5km every weekend to crafting really cool origami, the world is yours to explore if you give it some time to explore your passions and interests. Narrow your lists to the top 3 or top 5 and review it often. Use it as a note to yourself to indulge in your passions or when you're plotting your next career move.

CHAPTER 9
FIND YOUR HAPPY PLACE

According to a recent poll conducted in 2016 Harris Poll Happiness Index, only 31 percent of Americans considered themselves to be 'really happy' The reasons for not being happy are not just because of economic uncertainties or a tumultuous political climate. In most cases, it was the lack of optimism or the uncertainty of the future was what made people unhappy.

Dr. Jaime Kulaga, Ph. D, a licensed mental health counselor and life coach suggest that there are simple ways to improve your life instantly. Apart from the many exercises we've done in this book so far, here's another one you can try: Finding or Creating your Happy Place! A happy place, according to Dr. Jaime Kulaga is a space that you can go to find clarity and feel centered. When you step into a space that brings you peace or joy, it allows your mind to switch to destress mode so you can make more mindful decisions that are not swayed by emotions.

Before you start thinking about that island you want to go to as your happy place, STOP. Your happy place doesn't need to be some distant sunny island, nor does it even have to be real. The choice is yours really, to make it real or to imagine it in your mind.

It could be a bench in the garden, it could be a small nook in your favorite coffee place, a table by the window in the library, anything really. Your happy place can be an actual location or an imagined place that exists in your mind. It could even be a place you've visited before that boosts your mood in mere seconds. Or it could also be the swing in your childhood home.

The Journal of Experimental Psychology: General, published a study that showed how even thinking of traveling to a fantastical location and merely planning such a trip in your head can increase the happiness meter in your brain, more than reminiscing about the actual vacation afterward. This study suggests that imagining fantasy itself is a quick way to feel happy.

THE IMPORTANCE OF HAVING A HAPPY PLACE

The concept of a 'happy place' is an amazing one, where you can escape the stress and constraints of daily life and feel completely at ease. According to Dr. Kulaga, another benefit of a happy place is that it allows you to be present at the moment. By being present, we welcome our thoughts of gratitude.

Over time, gratitude increases our overall happiness as we begin to see the glass as half full as half empty. This will also then minimize our anger as well as anxieties. As we are all unique, we all have our own unique happy places. For some, a happy place doesn't necessarily mean a quiet space. It could be in the atrium of a busy shopping mall, seeing everyone walking up and down. For some, being in a club, dancing to music is a happy place whereas, for some people, it could just be in their bedroom, on a mat and just experiencing silence.

Having a happy place to be centered can help reduce mental issues such as anger, anxiety and even aggressiveness. There's also plenty of physical benefits such as better breathing, better body and mind connection. If your happy place is outdoors with nature, then you also have a healthy dose of vitamin D.

When we are acutely aware of our moments in life that provide us with happiness or pleasure, we also radiate this sense to maximize the amount of pleasure we get from them. According to the late Christopher Peterson, professor of Psychology, he describes a happy place as somewhere that is accessible easily, without penalty, neutral and it always contributes to the purpose of our lives.

Not all happy places have to be outdoors, not all happy places have to be indoors. Not all happy places have to be quiet or serene. In Shelley Levitt's book Live Happy, she profiled a man who had spinal muscular atrophy and is confined to a wheelchair since as a child. His happy place? The airport where he gets to see people coming and going, planes landing and taking off because it makes him feel that just by being there, he could be traveling somewhere new every time.

Your happy place is meant to give you a sense of pleasure and the freedom to lose ourselves in the moment. It is also a place that gives us the luxury of time to build relationships with our loved ones, the place to help us find meaning and it is also a place to achieve inspiration and our passions.

The wonderful thing about happiness is that it is subjective, and it is also a state of mind. If we have the right perspective, it is easier to find your happy place. So, the next time you find yourself in a place that makes you happy (for some it's even their own bedroom) take time to savor these moments. Write it down and tell yourself what you like this place and why you enjoy it so much, as a reminder in the event you need to conjure it up in your mind.

ESTABLISHING A HAPPY PLACE

Now that you know the importance of a happy place and what it could be, here are some tips to help you find your happy place:

- Recall places you've been where you appreciated the sounds

This could be the park you usually take your dog on a walk to or even the esplanade that overlooks the business center. It could even be your favorite pizza place that makes you feel warm inside. You can even make up a place with the sights and sounds you enjoy. Maybe you just like the sound of rain or the sounds of puppies in a pen?

- Summon the places where you've enjoyed the imagery

What colors speak out to you? Blue? Green? Black even? Think of the things that make you feel centered. Does looking at a mini fountain make you feel at ease? Or maybe sitting at the rooftop overlooking the city skyline?

- Choose a place where you can experience the elements that contribute to your happiness.

How does the smell of freshly cut grass make you feel? Or the smell of a cup of coffee brewing? Does a macaroni and cheese pasta make you feel warm inside? Or does the sound of the forest make you feel connected to yourself?

- Remember where you were when you experienced deep contentment and meaning

For Rebecca Bloomwood (from confessions of a shopaholic) her happy place was being in stores and doing some retail therapy. For friends in How I Met Your Mother, it was MacLaren's bar. For your colleague, sitting in a quiet meeting room to gather her thoughts is a happy place within her workspace. What about you? Think back to the places you've been to or imagine the spaces you feel like you can have deep contentment. Either create it in your mind or find it in the spaces you live in.

- Stay open-minded

Your happy place can be anything. As long as you do not confine your ideals and your thoughts into what a happy place SHOULD look like instead of what it is supposed to be for you, you can find peace and mindfulness anywhere you feel connected to the park, the store, the library, your kitchen even.

If you haven't found your happy place, this is a great time to start thinking about it just keep yourself open to any possibilities that make you feel contentment. Even looking at postcards or your favorite desktop wallpaper can be a source of inspiration for your happy place.

CHAPTER 10
ENHANCING YOUR EMOTIONAL INTELLIGENCE

Having or cultivating better emotional intelligence helps us work and relate better with the people around us. It also helps us have more meaningful relationships with our partners, our co-workers, friends, family and even acquaintances.

EMOTIONAL INTELLIGENCE AND SELF-LOVE

A better grasp of emotional intelligence will help us to monitor unpleasant feelings and you can also take actions to feel calm, happy and confident. Plenty of times, we leave it to external factors to bring us peace, joy and confidence but as we know, self-love is about loving ourselves unapologetically and knowing that happiness is our responsibility.

While it is great to celebrate all the positive achievements and events in our lives, we cannot depend on these events to elevate our happiness when we are angry, sad, upset or annoyed. Developing our Emotional Intelligence as part of our route towards self-love is a way of regulating our emotions so we can shift our mindset to be more in charge of our emotions even if the circumstances around us are negative. It also helps us relate to the people around us so that we can empathize with them and understand what they mean.

There are two kinds of Emotional Intelligence (EQ). The intrapersonal one and interpersonal kind.

1. Intrapersonal Intelligence: enables us to understand, regulate and acknowledge our own feelings, fears and motivations.

2. Interpersonal Intelligence: enables us to acknowledge, understand and regulate the emotions of others.

Being aware of these abilities means knowing that your emotions can drive our behavior and impact those around us, either positively or negatively. It also means we have the ability to manage these emotions, that of our own and that of others, especially at pressuring and stressful times.

Cultivating Emotional Intelligence

1. Observe your feelings

One of the first things we lose touch with is with our emotions especially when we focus all our energy into worrying about what to do next and what can be done better. Instead of focusing on our emotions, we instead choose to ignore them quite often. Things become worse when we start suppressing our emotions instead of dealing with it. When we keep covering up our emotions, we tend to lose control of them and that is not a good thing. When we experience an emotional reaction towards something, it is always because we have unresolved issues. The next time when you feel like a negative emotion is taking up space in your mind and heart, take a 5-minute breather, calm down and think about what you are experiencing and also the possible reasons that culminate these emotions. Write things down and try to identify your triggers and how you can deal with them.

2. Practice responding, not reacting

When we react, oftentimes we do it unconsciously in order to relieve the emotions that we are experiencing or also express what is going through our mind. When we respond in a conscious way, we are more adept at paying attention to our own feelings and we also become better and deciding how we will behave in reaction to these emotions and our feelings at that time. As we become more aware of our emotional triggers, we become more aware of how to adapt and how to respond as well as how to behave. For instance, if you know you get angry easily and you have a habit of throwing a temper especially when things get stressful, relive this moment when you are alone or when you are at home and think about how and what you would have preferred to react the next time to prevent yourself from experiencing the same trigger. Speak to your colleagues and tell them you need some time-out to gather your thoughts. Leave the room if you must, go out and get some fresh air. Count to ten even. Once you have calmed down a little, you will be better at dealing with the issues you face that made you angry in the first place.

3. Stay humble all the time.

Staying humble enables you to make better and meaningful relationships. When you presume that you are better than other people, it becomes harder to see your own faults and you will get emotional easily over things that do not meet your expectations or needs. To prevent this, you can start looking at things from a different perspective or even put yourself in the person's shoes and understand how they feel or how they would think of a certain situation. Doing so would make you more prone to understanding people's thoughts and feelings even more. You will also learn a thing or two about how to deal with situations that are similar.

Being humble is knowing that you are not any better than anyone else and also wise enough to know that you are special in your own way.

EXERCISES TO CULTIVATE EMOTIONAL INTELLIGENCE

ESTABLISHING YOUR EQ LEVEL

Emotionally intelligent people know what they want to achieve in life because they live and breathe this in and out. This does not only mean goals in career but in life, whether it is a happy marriage, a fulfilling career in a non-profit, feeding the hungry, becoming a humanitarian and son on. Start prepping yourself by writing your goals down, this has to everything that you want to accomplish in your life. Run a full marathon? Write it down! Scale the base camp of Mt. Everest? Write it down. Buy a house? Write it down. This will eventually turn into a long list so look at what you've written down and picked a dream or two that you feel most passionate about and apply the SMART goal setting approach which is:

- Specific

- Measurable

- Attainable

- Realistic

- Timely

You can create your vision board with the images and words that reflect your goals and place it where you can see every day. Include words and images that communicate how you feel about your most passionate goal. Break larger goals into actionable steps and create a plan to get started. Along the way, also recognize the people who play a big part in your life who can help you realize these goals and milestones. Acknowledge their support as this is a step towards emotional intelligence and also do not forget to thank them for the things that they do to help you get there.

- Exercise - Take an EQ Test

It is extremely helpful for you to take on an emotional intelligence appraisal. You can check one through TalentSmart. Doing an EQ appraisal can give you a deeper understanding of what your strengths and weaknesses are so that you can identify the areas that you need to focus on. This EQ appraisal that you do should not take be replaced with your daily reflection. This test is merely a jump start to strengthen your skills in EQ.

- Exercise- Make Use of Technology

Smartphones are another thing that has become so ingrained in our lives, many of us wouldn't know how to survive without your smartphones every day. There is an app for just about everything these days, even Emotional Intelligence, so why not make the most of the tools you have to help you establish a successful daily practice? There are several apps available that could help you enhance your sessions, with everything from timers to ambient sounds to help set the mood. If it helps make your daily practice more enjoyable, why not? You're more likely to stick to something if you like what you're doing.

IDENTIFYING YOUR SELF-AWARENESS
Becoming self-aware is part of becoming more emotionally conscious because when you are emotionally conscious, you are much more sensitive to the needs, emotions and state of the

people around you. One way of becoming more self-aware is to practice your reactions in a bad situation.

- Exercise- Negative Visualization

Negative visualization is the exercise that helps remind us of how fortunate we are. The idea here is simple. All of us need to do is to imagine that terrible things have taken place or, good things have not taken place. Keep in mind that this is not an exercise steering you into negativity. It is simply giving you an idea of the scale of catastrophe:

- You lost your wallet. But it could be worse - you could have lost all your possessions

- You are set up on a blind date. But it could be worse- you could have married without ever meeting your spouse beforehand.

- Your aunt met in an accident and is in the hospital recovering. It could be worse- she could have died.

- You met with an accident and fractured your arm. It could be worse- you could have lost a sense such as your sight or your hearing.

- You can also imagine how situations that you are about to embark on will go wrong.

This kind of therapy is not cultivating pessimism but instead, it makes you realize that things could get worse and things could turn bad, but it has not, and it did not happen to you.

In doing this exercise, you can try and imagine some catastrophes that could take place in the act that you are about to do or the situation that you are in. Maybe you could imagine having born in a time that you would miss having the convenience of the internet or being born at a time when women were not allowed to vote? Or traveling to far off places could only be done through sailing.

IDENTIFYING YOUR SELF-MANAGEMENT

From self-awareness, we now move on to self-management. What is self-management? With self-management, people are generally happier having control over their life and work. It is all about making the choices to do more than you are required to or more than you need to, and it is an excellent skill to build for life and work.

- Exercise- Count to 10

Counting to 10 has remarkable implications to your mind, body and soul. When we count to ten, it gives our minds a chance to figure out how to react so that we do not do things on impulse or say things that we would regret later. Whenever you feel angry, take deep breaths and count to ten in your head. This quick and short break will help you react in a more emotionally intelligent way rather and make the other person realize that not only are you upset but they would also calm down and offer to create a better way to manage the conflict. Do the same when you are stressed. Inhaling and exhaling out will help your mind and body relax and prevent you from getting a headache or getting an anxiety attack.

- Exercise- Set aside time for problem solving

If you are pointing out a problem, then you better be prepared to have a solution. Give constructive feedback to improve conditions and not worsen them. If you are going to point out a problem, then place as much effort into suggesting a good solution so you can offer yourself as a team member in solving an issue. Offer ways to make things better, not just point out mistakes and do not offer a helping hand.

A way to pay more attention to problem-solving is to mindfully spend at least 15 minutes a day to reflect on the issues you faced and how you could solve it or handle it better. This allows you to get to know your mind a little better and help you manage future situations in a way that solves conflicts and gets you and your team closer to a goal.

IDENTIFYING YOUR SOCIAL AWARENESS

With social awareness, you create and have the ability to comprehend and understand and also respond to the needs and emotions of others.

- Exercise: The Stripping Technique

Think of yourself as an onion and strip away the many layers that make you, you. Each layer that you build is years and years of living and the experiences that we go through. It is also the people we meet and the scenarios we encounter. It is not the situation itself but what each individual contributes to it. You need to stop considering your personal advantage or reputation may gain some insight into the scenario. You can ask yourself these questions:

What value does this situation bring to everyone? Most often, it is none.

What type of qualities does this situation require? If you have these qualities, wonderful but if not just view the situation without interfering as this would be a good chance for you to develop it.

For example, let us take a look at a common rite of passage for everyone, growing up and figuring out where we fit in the scheme of things.

As we grow up, we often struggle to find or decide what we want to do in life. Some people have it easy and know immediately what they like and what their passions are and then there are the rest of us who are still trying to find something fulfilling and meaningful to work towards.

Some of us have an excellent start in life while some of us end up doing something for the sake of money and some of us do it because it is expected of us or it's the only way we know.

Do you know the drill? Go to school, do well, get good grades, graduate, look for a job, find a spouse, get married, have kids and then work and work and work till we die?

Some of us fit into that mold while some of us do not. A good way to handle this exercise is to ask yourself 'What if money was not an issue?'. Whatever answer you give, think about where you are now and whether you are working towards this goal. This does not mean you need to quit your job but the exercise here is to incrementally work towards your goal, day by day, step by step.

IDENTIFYING YOUR RELATIONSHIP MANAGEMENT

The last aspect that you need to develop as part of your emotional intelligence is relationship management. Relationship management is the capacity to be aware of the emotions of the people that you communicate with and the emotions of your own to create a better and lasting relationship with friends, family and colleagues.

Relationship management is the analysis, identification and of course management of relationships with people who are inside and outside your social circle as well as the development through coaching and feedback. It incorporates your ability to persuade, communicate and lead all the while remaining direct and honest.

- Exercise: Tackle tough conversations

Nobody likes confrontations but it is an inevitable element especially where relationships are concerned. Avoiding critical discussions just results in no progress made. But as we know now, there are much better ways of handling things:

- When there is an argument, look at the things that you and the person you are arguing with and identify what are your common grounds.

- Get the person to share their side of the story or their point of view.

- Listen carefully to understand and not to answer. There is no point in going on the defensive immediately.

- Help the other person understand where you are coming from, describe your needs, your discomfort, your reasons and so on.

- Move the conversation forward by looking at your common grounds after you have discussed each other's views.

- Follow up if possible, with the person to see the progress based on what you have agreed upon.

While this is easier said than done, the methods described above can procure more positive and productive ways to handle conversations or arguments that are difficult. It helps you and the person listening.

Conclusion

Nourishing your body does not only mean consuming the right foods. It also relates to nourishing the mind. When the mind is nourished, the body will benefit physically, emotionally and mentally.

The path to self-love has no finish line. It is a continuous and adaptive process that begins first with self-awareness. The awareness that we need to change our ways and our thoughts, to be more compassionate and focused on our mental needs, to be more positive and to have better empathy.

When we work on ourselves, these benefits cascade to the people around us who will feel this new energy radiating from us. It will also affect our physical surroundings, the place we live in, the conditions we live in as well as the food we consume.

We begin to realize that we've not only been hurting our minds with negative and abusive thoughts, but we have also been hurting our bodies from the bad habits we've been practicing such as from excessive smoking and drinking, not getting enough sleep or exercise and accumulating clutter.

We began to peel off all of this unneeded and unwanted garbage and transform or reignite a better, more loving and compassionate person. Think of yourself as a phoenix, rising from the ashes.

As you get to the end of this book, remember that THIS IS NOT THE END. Your self-love journey has only just begun. You must continue practicing the various exercises and habits in order to make this second nature to you because self-love is an ever-growing and ever-changing process. This book helps you kickstart that process.

We wish you all the best and all the love in the world because you know what? You are worth it.

Printed in Great Britain
by Amazon